CW00868632

Discovering
Engineering
that changed the world

EARTHWORLD is a brand new imprint in the house of Veloce Publishing, showcasing innovative and thought-provoking books that are informative and entertaining. Produced to the same high quality of content and presentation as our existing books, EARTHWORLD is set to push the boundaries and expand horizons.

Also from Earthworld:

Mr Trump goes to Washington (Michael Mayor) – ***Political satire in 3D graphic novel form***

Coming soon ...

The Book of the Leica R-series Cameras (Brian Long) – ***The definitive history of Leica's SLRs***

Mum's not the word – Childless, Childfree (Denise Felkin) – ***A photographic essay challenging perceptions of women and motherhood***

Dairy Cows & Duck Races (Philip Dixon) – ***The tales of a young farmer***

www.veloce.co.uk

First published in September 2018 by Veloce Publishing Limited, Veloce House, Parkway Farm Business Park, Middle Farm Way, Poundbury, Dorchester DT1 3AR, England.
Tel +44 (0)1305 260068 / Fax 01305 250479 / e-mail info@veloce.co.uk / web www.veloce.co.uk or www.velocebooks.com.
ISBN: 978-1-787113-55-8 UPC: 6-36847-01355-4

Unless otherwise credited, all photography in this book is by Julian Edgar, using a Canon G15.

British Library Cataloguing in Publication Data – A catalogue record for this book is available from the British Library.
Typesetting, design and page make-up all by Veloce Publishing Ltd on Apple Mac. Printed in India by Replika Press.

Julian Edgar

Contents

Chapter 1: Rockets

Speed, and the way into space

I think I first heard of Peenemünde when I was about ten years old. I'd read a novel – *The End of Project 38* (by Edwin Johnson) – which was a World War II child's adventure story at its best. The book was published in 1963, the year I was born, but it was set 20 years earlier.

"From the heart of Nazi Germany, across the flat countryside and through the silent towns of occupied Belgium, rumble railway wagons loaded with long crates. Their contents are unknown, but terrifying rumours of a new secret weapon, a hideous menace to Britain's security, reach Intelligence chiefs in London."

I must admit I can remember nothing further of the plot of the novel – but I can remember those long crates on railway wagons. They contained rockets, the V2, and they came from a place that was hard to pronounce – Peenemünde.

As I became older, I read more and more about these rockets, finding a story of extraordinarily technological daring and discovery. In a time of enormous technological change, those missiles were near the pinnacle of World War II technical development. However, those advances were also being made within a dictatorial, murderous regime committing monstrous crimes against humanity: Hitler's Nazi Germany. And the V2 was as indicative of that regime as anything of the period – built by slaves dying in underground tunnels; developed by extraordinarily skilful engineers working in the most advanced research facility in the world.

So when we visited Germany, I very much wanted visit Peenemünde, the birthplace of space rocketry. But I also wanted to visit Mittelbau-Dora, the site of both a slave labour camp and the underground tunnels in which V2 rockets were built. I found both locations incredibly moving; the juxtaposition of technological achievement against amoral horror.

And then, when we visited the United States, I was fascinated to visit White Sands, the place where, after World War II, many of the German rocket scientists – and their missiles – ended up. For the US, turning a blind eye to the recent past, took full advantage of the spoils of victory, directly leveraging off the Nazi advances to kickstart their missile – and later, space – programs. At least at White Sands, the US fully acknowledge their technological debt, with a superbly restored V2 on display.

When as a child in primary school, I – and the rest of the school – were seated in front of a large B&W television to watch mankind landing on the moon, I had no idea that some of the very same engineers and scientists that had developed the German V2 were also responsible for the giant Saturn rocket that propelled the moon-landing men into space ...

The train to Peenemünde – the birthplace of modern rocketry.

V2 ROCKET

The world's first ballistic missile – at least 20 years ahead of the rest of the world

"We have invaded space with our rocket and for the first time – mark this well – have used space as a bridge between two points on the earth. We have proved rocket propulsion practicable for space travel. This third day of October, 1942, is the first of a new era of transportation, that of space travel," said General Walter Dornberger.

Devoted Nazi General Walter Dornberger and German rocket scientist Wernher von Braun were two men with a vision.

The drive, energy and resourcefulness of General Dornberger, mated with the ingenuity and passion of von Braun, created one of the greatest technological achievements of the last century. The fact that this achievement was the product of the ruthless regime of Nazi Germany and the fact that many thousands of people died as a result of its production and deployment, taint the image of what otherwise would be considered a monumental triumph.

The Nazi V2 rocket was the first modern rocket, and was decades ahead of every other country in the world. Every rocket that has followed owes a debt to its design. This rocket is at Peenemünde in Germany, at the site of the original research establishment.

World's first long-range ballistic missile

The V2 rocket was not only one of the most awesome weapons of World War II, it was also the first long-range ballistic missile to be actively used in combat. It provided technology that was directly responsible for the USA's success in the postwar space race. It was also so advanced that, immediately after the war, the US, Britain and the USSR had no idea how the rocket worked or how it was even fired ...

The V2 was the first unmanned guided ballistic missile. For guidance it used an advanced gyroscopic system that sent signals to aerodynamic steering tabs on the tail fins and vanes in the exhaust. It was propelled by alcohol and liquid oxygen fuel. The two liquids were delivered to the thrust chamber by two rotary pumps, driven by a steam turbine. The steam turbine operated at 5000rpm and was powered by two auxiliary fuels, namely hydrogen peroxide (80%) and a mixture of 66% sodium permanganate and 33% water.

This system generated about 25,000kg (55,000lb) of thrust at the start, which increased to 73,000kg (160,000lb) when the maximum speed was reached. The motor typically burned for 60 seconds, pushing the rocket to around 4800km/h (3000mph). The V2 rose to an altitude of 83-93km (52-58mi) and had a range of 320-360km (200-225mi).

Before launch, the empty V2 weighed 4500kg (10,000lb). It was then fuelled with 4900kg (10,0800lb) of 'A-Stoff' (liquid oxygen with a temperature of -183°C (-297°F), and 3710kg (8200lb) of a mixture of 75% ethyl alcohol and 25% water, called 'B-Stoff.'

The air batteries and nitrogen batteries were filled to 200 bar (2900psi) pressure, and after this process the rocket weighed 12,700kg (28,000lb). Electrical cables were then connected and gyroscopes powered-up by 28 volts/60 amps, DC. The cables remained connected until launch, batteries taking care of power supply during flight.

After everything was set, pressurised air (at 32 bar (460psi)) pushed the hydrogen peroxide and the sodium permanganate into the 430kW (580hp) turbine. This caused the turbine to rotate at 3800rpm. This turbine

The power station is the only large building that remains at Peenemünde – the laboratories, supersonic wind tunnels and accommodation now all gone. The building houses an excellent, thoughtful and nuanced museum.

A fully-restored, cutaway V2 at White Sands Missile Range, New Mexico, USA. German scientists and their missiles were brought to the US immediately after World War II, and both played a large part in the development of US missiles and the rockets that powered the space program. White Sands has an excellent visitors' centre.

powered two pumps that injected the ethyl alcohol at 23 bar (330psi) pressure via 1224 nozzles (58kg/sec (130lb/sec)) and liquid oxygen at 17.5 bar (250psi) via 2160 nozzles (72kg/sec (160lb/sec)) into the burn chamber at 23 bar (330psi).

The mixture ignited, whereupon a temperature of 2500°C (4500°F) at 15 bar (220psi) pressure was reached – but this was not enough to lift the V2. After checking to ensure the propulsion was working properly, the burn speed was then raised, and cables electromagnetically disconnected.

The V2 engine burn chamber temperature was about 2700°C (4900°F). The chamber wall was cooled by the liquid ethyl alcohol flowing via the double wall of the beam tube and burning chamber, which also resulted in the heating up of the ethyl alcohol.

During burn time, the V2 was steered by four carbon graphite rudders (in the engine exhaust) and four vanes (at the fins), controlled by three gyroscopes.

Impact speed of over 3000km/h

This huge German rocket – over 14 metres (46ft) long – hurtled a 975kg (2100lb) warhead over 80km (50mi) high, and hundreds of kilometres down-range to its target. After engine shut-down, the V2 flew to a height of 97km (60mi), and then fell to earth with an impact speed of 3240-3600km/h (2000-2200mph).

When launched against targets close to the operational range of the vehicle, the deviation between target and impact was normally 7-17km (4-10mi). This made the rocket suitable only for use against widely populated areas. At shorter targeting ranges, the accuracy of the V2 was improved. The Leitstrahlstellung was a 'guiding beam' that improved accuracy of the V2 somewhat, during the later days of the campaign. One-quarter of all V2 rockets were guided with the Leitstrahlstellung.

A V2 impact would be as follows. First, a whip-cracking sound of a blast wave created by the rocket (moving far faster than the speed of sound) bounced off the point of impact just split seconds before the flash of impact. This was followed by the chaos of the explosion, with debris and earth churned skyward. Soon, there followed the whine and rush of whistling air, as the sound caught up with the rocket, followed by a deafening roar of the incoming rocket, which tapered off to silence.

There could be no audible warning; the V2 impacted at three times the speed of sound ...

Since the V2 was not operational until late 1944, the countless funds, materials, and manpower that were used in its construction could have been better used to produce more planes and tanks. It was purely a 'Vengeance Weapon,' but there was no countermeasure that the Allies had to stop it. The V2 offensive would last from September of 1944 until March of 1945, with over 3000 rockets launched in this time period. The London area was hit by over 500 rockets, and several hundred more dropped in surrounding counties. At first, London and Antwerp were the primary targets, but rockets also fell around Ipswich and Norwich, and many Allied-held targets in France, Belgium and Holland, and even on Germany itself.

Built by slaves working underground

Despite the fact that the V2 was a weapon of war, more slave labourers died building the rockets than people who were struck down by the explosive warhead that the V2 carried.

From 1943, the V2 rockets were constructed in an underground system of tunnels under Kohnstein Mountain, near Nordhausen. This move was as a response to Allied bombing of existing works – an environment protected from bombing needed to be found. The new plant became known as Mittelwerk.

After meeting with Hitler on August 18th, SS Chief Heinrich Himmler had informed Armaments Minister Speer that he was personally taking over V2 production and placing SS Brigadier General Hans Kammler in charge of the complex. It was Kammler who had been in charge of building the infamous extermination camps and gas chambers at Auschwitz-Birkenau, Maidenek, and Belzec.

On August 28, 1943, two days after the choice of Mittelwerk, the SS delivered the first truckloads of prisoners from the concentration camp at Buchenwald to begin the heavy labour of expanding and completing the tunnel system. Dora was the name given to the Buchenwald sub-camp.

Along with jet engines and the V1 cruise missile, the Mittelwerk factory produced some 4575 V2s between August 1944 and March 1945.

Prisoners were divided into two groups of workers: Transport Columns and Specialists. The former did the often backbreaking work of manually transporting much of the material that entered or left the tunnels, while the latter carried out more skilled assembly and testing work. Detainees working in the tunnels were divided into day and night shifts, each working for 12 hours straight. Every four weeks, the workers changed shifts. Each prisoner work group, or 'kommando,' was headed by a prisoner leader (Kapo).

Teams of six transport prisoners were assigned to carry into the tunnels the empty aluminium tanks for the rocket from the outside storage depots. Designed to be lightweight for their size, each tank still weighed about 150kg (330lb) – or about 25kg (55lb) per worker. The workers formed two parallel columns and grasped the hand of their counterpart alongside. The tank was then

Underground in one of the original tunnels in which V2 missiles were built by slave labour. The Nazis established a camp – Dora – to supply the labour. Conditions in both the tunnel and camp were horrendous. Each completed V2 missile cost an estimated six terrible deaths.

Despite the time that has passed, and the fact that the US, Great Britain and the USSR all combed these tunnels for parts to help in their respective rocket development programs, V2 components still litter the tunnels of the underground factory. Tours are available (in German only). It is a daunting, terrible place.

In addition to the use of liquid fuel, the V2 was advanced in its control and guidance system. Pictured here in London's Science Museum is part of the gyroscopic control system. Missile guidance systems of the 1950s and 1960s bear an uncanny resemblance to this.

Some of the extraordinary engineering of the V2 can be seen below in this cutaway view of the turbine fuel pump. It operated at 5000rpm and was powered by steam generated by two auxiliary fuels. Photographed at the Deutsches Technikmuseum in Berlin.

slung on their joined arms. If a group dropped its tank (not uncommon, since these skeletons of men were often already weak and sick), the SS guards and Kapo were there to kick and beat them with truncheons until they could lift their burden and continue once again. Since much of this work was done in the dead of one of the coldest winters on record, the workers were usually slogging though snow, ice, or freezing rain and mud. It is hard to imagine what this must have been like. On their feet they wore wooden clogs, and had very little protection from the elements.

It is estimated that of the 60,000+ detainees employed in and around the Mittelbau complex over a 20-month period, 26,500 did not survive. One author attributes 15,500 of these deaths to the camps or to 'transports,' and 11,000 to the period in April 1945 when the camps were evacuated by the SS in the face of

The crematorium is the only original building that survives at Dora labour camp. A powerful museum is located at the site, and together with a tour of the tunnels in which the V2s were built, is an incredibly sobering experience.

Rusting in the rain – the rear of a V2 and part of its engine. New Mexico Museum of Space History, USA.

If you'd like a piece of a V2, you can buy it. Incredibly, genuine V2 parts are advertised quite frequently on eBay – here is part of a V2 rocket engine that I have in my collection.

Rocket engines have been fitted to cars as well as space projectiles and aircraft. Here are two of them.

The Opel Rak 2 was a rocket-powered car built in Germany. On May 23, 1928, it reached a speed of 230km/h (143mph), propelled by 24 solid-fuel rockets. The vehicle was driven by Fritz von Opel, of the Opel car company.

The Opel Rak 2 of 1928. It used 24 sold-fuel rockets and reached a speed of 230km/h (143mph). This re-creation is at the Technik Museum Speyer, Germany.

Curiously enough, the number of people exploring rocketry in Germany in the 1930s was so small that the car has a direct link to the post-WWII American space program. One person who worked on the rocket car was Max Valier, who was later a member of the same amateur rocket society that attracted members such as Wernher von Braun and Arthur Rudolph, two of the men who worked on both the V2 rocket and then the US Saturn rocket that took men to the moon. Valier died in 1930 when a rocket engine he was working on exploded.

Blue Flame – the car that set the world Land Speed Record in 1970. It's now in the Technik Museum Sinsheim, Germany, where I took this picture of the car and my son, Alexander.

The Blue Flame was a rocket car of a very different type. The vehicle achieved the world land speed record on Bonneville Salt Flats in Utah on October 23, 1970. It averaged 622.407mph (1,001.667km/h) for the flying mile and was driven by Gary Gabelich. It was 11.3m (37ft) long, weighed 2.2t (4950lb) and was powered by a liquid natural gas-hydrogen peroxide rocket engine. Thrust was said to be as high as 10,000kg (22,500lb) but the engine ran at much less than this during the record run. (Later, an estranged member of the design team, Ray Dausman, implied the lower thrust was because the remaining members of the team didn't know what they were doing – see the book *The Reluctant Rocketman* by Dausman's daughter, Sarah Kasprowicz.)

the American advance. This evacuation was especially barbaric. The SS shot prisoners, herded them into barns and burned them alive, left them to die if they were too sick to walk, or made them part of walking or rail convoys headed to other concentration camps.

Each operational V2 to come off the Mittelwerk line cost an estimated six terrible deaths.

MESSERSCHMITT ME 163 KOMET

Swept wings and propelled by a rocket engine that used flesh-dissolving fuel!

Nazi Germany didn't just exploit rocket propulsion for development of the extraordinary V2 missile – it also used it to develop the fastest aircraft in the world.

The Messerschmitt Me 163 Komet of World War II is one of the most remarkable aircraft ever produced. Not only was it the first (and only) tailless, rocket-powered aircraft to see service, but it also held the unofficial aircraft speed record, and could climb at a simply stunning rate. Designed to shoot down Allied bombers, the aircraft had sufficient climb performance to be scrambled from base when the high-altitude bombers were almost overhead.

But it must be said, the Me 163 also had some less pleasant aspects – like fuel that in an accident could literally dissolve its pilot, a total engine operating time per sortie of under 10 minutes, and the requirement for a very high landing speed ... while unpowered.

Two technological breakthroughs

The genesis of the Me 163 was predicated on two technological breakthroughs – and, of course, the political environment of Nazi Germany where high technology military developments were well-funded.

The first technological development was in the development of radical, tailless aircraft. Following World War I, Germany was prevented from developing powered aircraft that could be used for military purposes. However, designing and building gliders was permitted, and the best of the contemporary German aircraft designers (and, subsequently, German pilots) were involved with gliders. One such designer was Alexander Martin Lippisch. Born in Munich on November 2, 1894, Lippisch is said to have been inspired to follow an aviation career after seeing a flying demonstration by Orville Wright in Berlin in 1909. He first designed a tailless aircraft in 1921, and throughout

The Me 163B in the Science Museum, London. The tiny propeller on the nose drives a generator, not the aircraft, which in fact is propelled by a powerful rocket engine. The aircraft took off on a trolley that was dropped; it landed on the skid that can be seen deployed here.

the 1920s and the 1930s produced some 50 tailless aircraft designs. (He also developed delta wing designs – his first tailless, delta wing aircraft flying in 1931.)

The other significant technological development of the era was in rocketry. In the mid-1930s, Hellmuth Walter left his position at the German Army Ordnance Department and set himself up in business. Walter's main business interest was in the development of rockets. Funded by all three German armed services, by 1936 he was developing rockets for aircraft propulsion.

Two types of rocket engine were produced – they could be differentiated into 'cold' and 'hot' types, the nomenclature depending on the chamber temperature achieved during the burn.

The Walter RI-203 development model was a 400kg (882lb) thrust engine fuelled by a mixture of T-Stoff (80 per cent hydrogen peroxide with oxyquinoline or phosphate as a stabiliser, and 20 per cent water) and Z-Stoff (an aqueous solution of calcium permanganate). This engine was experimentally flown in a Heinkel He176 V1 on June 20, 1939.

The He176 was a single seat, pressurised aircraft with small elliptical wings. It was the world's first aircraft designed from scratch to be powered by a rocket. However, performance was below expectations. The Reich Air Ministry then ordered the German Research Institute for Gliding Flight (DFS) to produce a new aircraft based on the DFS 39, a tailless aircraft designed by Alexander Lippisch. This new aircraft was to use the Walter rocket motor. DFS 39 had wing-tip mounted rudders, which Lippisch realised would likely flutter, and that a central fin and rudder would give better control. So was born the DFS 194.

Initially known only as Design X (the '194' designation came later), the aircraft was developed under very tight security. In fact, security was so strict that Lippisch and his team had no detailed data on the rocket engine that was to propel the aircraft! They knew only general information on the motor's size, weight and thrust specification, but had no detailed drawings.

The DFS facilities proved to be ill-equipped to build the metal fuselage of the 194, so Lippisch and his team transferred in January 1939 to the Messerschmitt company. The airframe was completed in 1940 and sent to Peenemünde West airfield for installation of a Walter RI-203 engine. Ground runs of the aircraft commenced in October 1939, but the engine installation, where the engine was integrated with the fuselage (only the fuel and compressor air-lines could be easily removed), was found to be unsound. The aircraft never flew under rocket power; however, gliding tests were carried out.

Despite these unimpressive beginnings, it was decided that three development prototypes be constructed – these

The Me 163's rocket engine. Its fuel was so corrosive that it could dissolve human flesh – an unpleasant end for at least one pilot. Photographed at the Deutsches Technikmuseum in Berlin, Germany.

were later officially designated as the Messerschmitt Me 163A, Versions 1-3.

Me 163A

The Me 163A V1 was completed during early 1941. Unpowered flight testing commenced in the spring of 1941, before the aircraft was shipped to Peenemünde West for installation of a 750kg (1653lb) thrust Walter RII-203 rocket motor. The new rocket motor was easily removable from the airframe, and could be manually regulated in thrust by the pilot over the range of 150kg (331lb) to full thrust. Furthermore, the engine could be cut and then relit by the pilot while in flight. But the rocket engine proved to be erratic in its performance – explosions and accidents occurred.

However, on October 2, 1941, an Me 163A was towed to an altitude of 4000m (13,000ft). There, the pilot started the rocket motor and accelerated. Observers on the ground, watching through binoculars and other visual tracking instruments, suddenly saw the long black smoke trail from the rocket engine waver – and then stop. The pilot, travelling at just under 1000km/h (600mph), had lost control of the tiny aircraft. Shutting off the engine, the pilot slowly gained control, and was then able to glide back to base to make a normal landing.

The loss of control had occurred as the aircraft ran into compressibility effects – the change in the way that air behaves when nearing the speed of sound. The loss of control was of course incomprehensible at the time – the Me 163A had just gone faster than any other aircraft, and little was known of the problems of compressibility. Recording instruments in the aircraft showed that it had

ROBERT GODDARD – AMERICA'S ROCKET PIONEER

It's true that Germany led the world in rocket development, however, it was not a German scientist but instead an American, Robert Goddard, who constructed and successfully tested the first rocket using liquid fuel. As early as 1920 he suggested using rocket propulsion to get to the moon, and he discussed the use of gyroscopes, steering by vanes in the rocket exhaust, power-driven fuel pumps and other devices. Progress on his work, entitled 'Liquid Propellant Rocket Development,' was published by the Smithsonian in 1936 – and it was a document read closely by the Germans. However, the German authorities were prepared to invest substantially in rocketry; the same support was not available in the US.

A rocket motor produced by Robert Goddard in the US, probably for his first rocket launch in 1930. The motor is on display at the Steven F Udvar-Hazy Museum, Virginia, USA.

reached 1004km/h (624mph) – a world record. In war time this was, of course, kept secret – the Allies having no aircraft that could travel even remotely as fast.

Me 163B

Based on the performance of the Me 163A, it was decided that a new model of Me 163 aircraft be developed – the B series. While appearing similar in many respects, these new aircraft represented almost a complete redesign.

To hold more fuel, the fuselage was larger. A small propeller was added to the front to power a generator for the 24V electrics. A much more powerful rocket motor was also fitted. The Walter 109-509A motor produced no less than 1500kg (3300lb) of thrust, and used a more volatile fuel mixture of T-Stoff (80 per cent hydrogen peroxide and 20 per cent water) and C-Stoff (hydrazine hydrate, methyl alcohol, and water).

However, the motor was again quite unreliable. Aircraft could explode on engine start-up, the engine could cut out when experiencing even fractional negative 'g,' and the engines sometimes stopped just as the aircraft were getting airborne. Furthermore, the aircraft took off on a wheeled trolley that was then dropped. If dropped at the wrong time, the trolley could bounce back up into the air, hitting the aircraft. Landing was on a skid, and heavy landings caused pilot injuries.

The fuel was so dangerous that pilots and ground crew wore protective suits, but in a crash these were of little use. In one crash, in which the pilot died, parts of his head and arm were literally dissolved away by the fuel.

330km/h faster than a Spitfire

However, once in the air, and with the rocket motor operating, the Me 163 B had extraordinary performance. Including take-off time, the Me 163 B could reach 9800m (32,000ft) in an incredible 3 minutes and 19 seconds! To reach 12,200m (40,000ft) took just 3 minutes and 45 seconds. One source quotes an initial climb rate of 4900m (16,000ft) per minute. Compare that with the Spitfire, that had an initial rate of climb of just 1400m (4700ft) per minute. Maximum service speed of the Me 163B: in the region of 960km/h (600mph); the Spitfire: 630km/h (390mph). (And outright maximum speed of the 163B? One source says that in 1944 the aircraft reached 1130km/h – 700mph.) However, the Me 163B had a practical range of only about 130km (80mi), much of that unpowered (the aircraft would glide back to its airfield). The maximum duration of rocket power was only 7 minutes and 30 seconds.

So why develop such a sophisticated aircraft with such a short range? Was the decision to go ahead with the development of the Me 163B fatally flawed? The answer is 'yes' – but only in retrospect. At the time, the Walter company was promising a rocket motor with much greater duration – in fact, a motor firing time that would have made the aircraft far more practical in a strategic sense. Furthermore, the relative speed of the Me 163B, versus the bombers it was meant to shoot down, required the development of new gun-aiming devices. Without these developments, the window of time in which the pilot could manually fire the guns was too short for consistently successful results.

To put this another way: the war ended before the aircraft could be developed to its full potential.

Ground-level view of the 111m (363ft) tall Apollo 11 Saturn V on its way to the launch pad, May 20, 1969. Note the size of the cars. (Courtesy NASA)

SATURN V MOON ROCKET

In terms of height, weight and payload, it remains the largest rocket ever built

The Saturn V carried aloft the 45-ton Apollo Spacecraft on earth orbital and lunar missions from 1967 to 1972. It also launched the 120-ton Skylab into earth orbit on May 14, 1973. So how did the rocket come about?

Beginnings

On April 11, 1961, Major Yuri Gagarin made his historic

Birds scatter as the mighty Saturn V rocket lifts off, with Apollo 4 atop. (Courtesy NASA)

Just one the cluster of five F-1 rocket engines on the first stage of the Saturn 5 rocket. Each F-1 engine could develop 690,000kg (1,522,000lb) of thrust. On display at New Mexico Museum of Space History, USA.

voyage around the world in space. The Soviet Union and the United States then became involved in a technical race for prominence in space. On May 25, 1961, President John F Kennedy addressed Congress. His speech launched an historic challenge to technology:

"Now it is time to take longer strides ... time for this nation to take a clearly leading role in space achievement which in many ways may hold the key to our future on earth ... I believe that this nation should commit itself to achieving the goal, before this decade is out, of landing a man on the moon and returning him safely ... In a very real sense, it will not be one man going to the moon, it will be an entire nation."

When the decision was made to undertake a manned lunar landing effort, there was no rocket in the country even approaching the needed capability.

On January 10, 1962, the National Aeronautics and Space Administration announced that it would develop a new rocket, much larger than any previously attempted. It would be based on the F-1 rocket engine, the development of which had been underway since 1958, and on the hydrogen-fuelled J-2 engine, upon which work had begun in 1960.

The Saturn V program was the biggest rocket effort undertaken in the United States. Its total cost was above $7 billion. The rocket consisted of three stages and an instrument unit, it was 111m (363ft) tall and weighed approximately 2800t (6.2 million lb) when fully loaded.

Developmental work through the 1960s culminated in the first launch of the Saturn V on November 9, 1967. On December 21, 1968, the first manned flight (and the third launch) of the Saturn V took place with Borman, Lovell, and Anders in the Apollo 8. Two more flights carried men both around the earth and around the moon to test the hardware for the attempt at a manned moon landing.

In July of 1969, the sixth launch of the Saturn V carried the Apollo XI mission of Armstrong, Collins, and Aldrin to the first manned landing on the moon. Six additional launches succeeded in achieving five more lunar landings up to December 1972. The last launch of the Saturn V orbited the unmanned Skylab Workshop.

First stage

The S-IC stage provided the first stage boost of the Saturn V launch vehicle to an altitude of about 61,000m (200,000ft) and accelerated the vehicle velocity to 7700ft/sec (2300m/sec). It was 300,000lb (136t) in weight, 33ft (10m) in diameter and 138ft (42m) long. It was powered by five F-1 engines generating 7.5 million lb (3.4 million kg) total thrust. The booster burned 203,000 gallons (768,000 litres) of refined kerosene, and 33,000 gallons (12,500 litres) of liquid oxygen in 2.5 minutes, its total burn time.

Second stage

The S-II stage provided the second stage boost for the Saturn V. This stage was powered by five J-2 engines that generated a total thrust of a million pounds (450,000kg). It was 33ft (10m) in diameter and weighed 95,000lb (43t) empty and more than a million pounds (450t) when loaded. It burned 260,000 gallons (nearly a million litres) of liquid hydrogen and 83,000 gallons (300,000 litres) of liquid oxygen during its six-minute flight. At engine cut-off, the S-II stage separated, and re-entered the atmosphere where it disintegrated.

Third stage

The S-IVB was the third booster stage. Its single J-2 engine was designed to boost the payload into a circular orbit on the first burn, then with a second burn, boost the payload to a proper position and velocity for lunar intercept. This stage weighed approximately 34,000lb (15t) dry. The vehicle was 21ft 8in (6.6m) in diameter, and 58ft 6in (17.8m) long. The typical burn time was 2.75 minutes for the first burn, and 5.2 minutes for the second.

Instrument unit

The vehicle instrument unit sat atop the third stage. This unit, which weighed approximately 4500lb (2t), contained the electronic gear that controlled engine ignition and cut-off. The IU contained the guidance, navigation, and control equipment which guided the vehicle through its earth orbit and subsequently into its mission trajectory. Diameter of the IU was 21ft 8in (6.6m) and the height was 3ft (1m). Above the instrument unit sat the spacecraft.

The Saturn V was the first large vehicle in the US space program to be conceived and developed for a specific purpose. It remains to this day an enormously impressive technological achievement.

Acknowledgement: Some of the text of this chapter is based on material drawn from www.v2rocket.com (The V2 Rocket) and the American Society of Mechanical Engineers (The Saturn Moon Rocket), and is used with permission. The Saturn rocket is a Historic Mechanical Engineering Landmark.

Chapter 2:

Inventing flight

Two geniuses and their incredible aircraft

Orville and Wilbur Wright were genuine engineering heroes. Despite their relatively humble beginnings, these men were the first to ever build – and then successfully fly – a powered aircraft.

I've often read descriptions of their brilliance that damns them with faint praise: they were 'just bicycle mechanics,' their work built heavily on the efforts of others, and so on. In fact, they were simply brilliant engineers, with an astounding work ethic, and the ability to both physically make things and also theorise about outcomes.

In case you think I overstate the brothers' abilities, consider these points. They built their own wind tunnel, and tested in it almost 200 wing sections; they built an internal combustion engine with the then best power/weight ratio of any engine in the world; they made every part of their own aircraft – from that engine through to propellers, to wing and control systems; they developed the concept of an aircraft banking into turns – and a lot more.

Going to the site of their first flight in the sand dunes of Kill Devil Hills was one of the most moving experiences of my life, and seeing the original 1903 Flyer in the National Air and Space Museum in Washington DC was extraordinary.

This chapter takes the story beyond that first aircraft to the Wright Flyer III, the world's first practical aircraft.

Probably the most famous technical photo of all time.

THE WRIGHT FLYERS

Step by step development of a true flying machine

Humans have dreamed of flight among the birds since ancient times. The mythical story of Daedalus and Icarus escaping the maze on wings of feathers and beeswax is perhaps the best known of many early legends involving flight. Leonardo da Vinci's concepts for an ornithopter and protohelicopter show that the dream remained alive during the Renaissance. These ideas for human-powered flight were impractical, but the interest never waned.

During the first half of the 19th century, Englishman George Cayley began serious work that he hoped would result in an airplane. He succeeded in building a glider large enough to carry his coachman on a flight in 1853 – after which the coachman promptly resigned!

But in the 19th century, other experimenters, notably John Montgomery, Octave Chanute, and Otto Lilienthal, built and flew a variety of gliders. Hiram Maxim and Samuel Langley were the most notable of those who attempted to marry a glider to a steam engine, the most reliable power plant of the day. While both performed tests that suggested a successful, manned flight was imminent, neither met with success. The spectacular failures of Langley's Great Aerodrome on October 7 and December 8, 1903, were recorded in widely circulated photographs.

It remained for two unassuming brothers, Wilbur (1867-1912) and Orville (1871-1948) Wright, bicycle mechanics from Dayton, Ohio, to methodically work through the

Visiting the site of the first flight at Kill Devil Hills, North Carolina, was an extraordinary experience. (Courtesy Georgina Edgar)

problems associated with flight – lift, power, structural strength, and control – and produce a machine that could accomplish the task.

For four years, the brothers systematically tackled these problems, and overcame them one by one. Along the way, they provided a model for aeronautical research and generated information that would help form the foundation for what we now call aeronautical engineering.

First flight

There is no more famous event in aviation history than the first powered flight by Orville Wright on December 17, 1903. This 12-second flight covering 36.6m (120ft), marked the first time a human had successfully piloted a self-propelled machine, that rose into the air on its own power and landed on ground as high as that from which it had taken off.

Fortunately, the Wrights had enlisted the aid of some nearby lifesavers at Kitty Hawk, and one of them, John Daniels, snapped a camera shutter as the Wright Flyer rose into the air, taking one of the most famous photographs of all time. Orville Wright made one more flight that day, and his brother Wilbur made two, with his last one covering 260m (852ft) in 59 seconds. Shortly thereafter, the brothers' jubilation at their success was dampened when a gust of wind picked up the fragile plane and flipped it over, causing major damage and bringing the first series of flight tests to an end. Between them, the Wrights had amassed a total of just 98 seconds in the air.

The significance of these first flights cannot, and should not, be minimised. They were, after all, the achievement of the lofty goal that many great minds had sought in vain for centuries. The Wrights, however, were practical men, and even while announcing their success and taking preliminary steps to protect their invention, they realised that their quest was far from over. The Wright Flyer had indeed flown, but it had flown basically in a straight line and only a few feet above the ground. This was far short of the capabilities needed in a practical, useful airplane.

To be truly practical, an airplane would have to be able to climb to an altitude that would clear trees and buildings, and it would need to be fully manoeuvrable so that a pilot could turn it and fly in any direction at will. In addition, a practical airplane would have to be reasonably safe and easy to control.

At the Wright Brothers National Memorial, Kill Devil Hills, where they first flew. The inscription on the memorial reads: "In commemoration of the conquest of the air by the brothers Wilbur and Orville Wright, conceived by genius, achieved by dauntless resolution and unconquerable faith."

Returning to their home in Dayton in time for Christmas, the Wrights realised that they were at a crossroads. The bicycle shop was their livelihood; they had been pursuing their dream of flight as a sideline. Now that they had achieved what they thought would be their goal, they recognised how much development work still remained to be done. If they were going to undertake that – and to realise any financial reward for their work – the brothers would have to make a full-time commitment to aviation.

They decided to turn the shop operations over to Charles Taylor (1869-1956), the machinist who had built the Wright Flyer's 9kW (12hp) gasoline engine, and devote all their time to perfecting their airplane. In addition to managing the shop, Taylor would continue to contribute his talents to flying machine development as needed.

The Wrights also realised that they needed a location close to Dayton where they could fly on a regular basis to test new ideas. A 34 hectare (84 acre) field called Huffman Prairie, about 13km (8mi) east of town fitted the bill. It was easily accessible, yet somewhat isolated from prying eyes, and it was large enough to suit their needs. Its owner, Torrence Huffman, was sceptical of the Wrights' endeavours, but he granted them permission to use his field.

By April 1904 they had cleared the field, erected a wooden shed, and begun assembling their next flying machine, now known as the 1904 Wright Flyer II.

The second Wright Flyer was almost a duplicate of the 1903 Wright Flyer. Accordingly, it suffered from all the same problems, especially those of power and control. While the first Wright Flyer had flown, it was very difficult to control. The first four flights had been rather wobbly ventures, with the craft tending to pitch up and down through the air. In fact, Wilbur's first attempt at powered flight on December 14, 1903, ended almost before it began when the nose pitched up and caused the plane to stall on take-off. The brothers had made no attempt to turn in the air. It was all they could do to keep the craft aloft and make reasonably controlled landings.

Nevertheless, the original Wright Flyer's design was the

Re-creation of the world's first aircraft hangars. Kill Devil Hills.

Inside one of the re-created Kill Devil Hills hangars.

The 1903 Flyer in the National Air and Space Museum, Washington DC. Orville Wright originally donated the aircraft to the Science Museum in London, rather than allowing the Smithsonian in the US to display it. The reason? The Smithsonian had exhibited Samuel Langley's Aerodrome with a misleading label stating that it was the "first man-carrying aeroplane in the history of the world capable of sustained free flight." Langley had been Secretary of the Smithsonian. The Wright Flyer came back to the US only after Orville died in 1948.

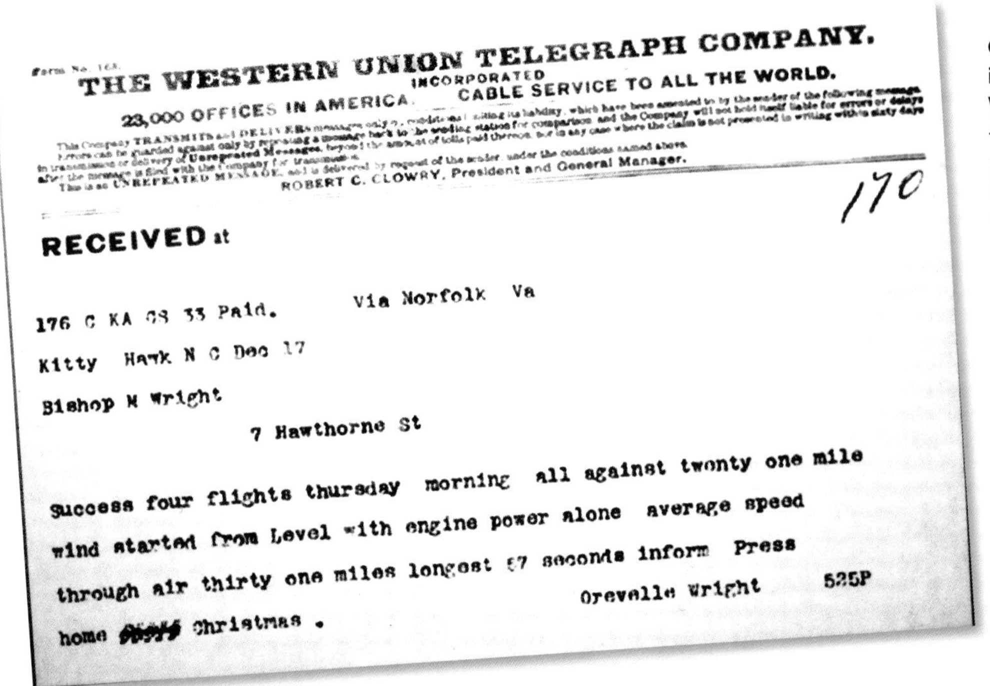
Form No. 168.

THE WESTERN UNION TELEGRAPH COMPANY.
INCORPORATED
23,000 OFFICES IN AMERICA. CABLE SERVICE TO ALL THE WORLD.

ROBERT C. CLOWRY, President and General Manager.

170

RECEIVED at

176 C KA CS 33 Paid. Via Norfolk Va

Kitty Hawk N C Dec 17

Bishop M Wright

7 Hawthorne St

Success four flights thursday morning all against twenty one mile wind started from Level with engine power alone average speed through air thirty one miles longest 57 seconds inform Press home Christmas . Orevelle Wright 525P

One of the most famous telegrams in history. As instructed, Bishop Wright – the brothers' father – informed the press, but no-one took much notice. It would be years before the world understood the significance of what the brothers had achieved. Photographed at the National Air and Space Museum, Washington DC.

only logical starting point. To replace the damaged 1903 engine, the Wrights designed, and Taylor built, a new engine that produced about 3kW (4hp) more power, and the brothers relocated the elevator's pivot point to reduce pitch sensitivity. The gas tank and radiator were moved toward the rear. They also flattened the wings' camber from 1⁄20 of the chord to 1⁄25 (a mistake that reduced lift) and strengthened the rudder structure.

Wright Flyer II was ready to fly on May 23, 1904, but it failed to leave the ground in its first test – in front of about 40 unimpressed onlookers. Three days later, the frail craft managed to fly a brief, 7.6m (25ft) hop. The brothers tried 40 more flights through August, but the longest one covered only slightly more than three-quarters the distance of the longest flight at Kitty Hawk. The problem was a lack of wind. They had launched into a strong headwind at Kitty Hawk, but the air over Huffman Prairie was calm, sometimes gusty and less dense. The plane needed an air speed (the relative speed through the air) of at least 37km/h (23mph) to lift off. That meant a ground speed of less than 8km/h (5mph) into the Kitty Hawk wind, but ground speed and airspeed were essentially the same in Dayton. The Wright Flyer simply could not reach the necessary speed by itself before reaching the end of the launching rail.

The answer was a catapult to accelerate the plane down the rail. The brothers built a 6m (20ft) tower at one end of the rail. A 725kg (1600lb) weight was suspended from its top by a rope that ran through pulleys to the far end of the rail, and back to the small launching carriage. When a hold-down rope was released, the falling weight provided the extra boost necessary to achieve more than the minimum lift-off speed. The catapult's additional force was great enough that the rail could be shortened to only 18m (60ft) in length.

The first test of the new launch system on September 7 was a rousing success. Once the Wright Flyer II was safely in the air, it performed reasonably well, and flight distances and complexities increased. By September 15, the Wrights were making flights up to a 800m (half a mile) in length, and they could make complete turns in the air. Orville flew a complete loop around the field on September 20, taking about a minute and a half to fly over 1.2km (4000ft). While they achieved progressively longer flights, accidents were common. Fortunately, most were minor, but the brothers suffered a variety of bumps and bruises, and made countless repairs to the machine.

As the brothers' piloting experience grew, they realised that Wright Flyer II was next to uncontrollable. Soft landings were rare. The biggest problems involved pitch. The craft would often begin to oscillate vertically during straight and level flight for no apparent reason, frequently resulting in a rough landing. Moving the centre of mass forward by attaching a weight under the elevator helped some, but clearly more needed to be done to get pitching performance under control. Wing warping controlled roll very well, but the plane would tend to lose altitude when banked into a turn. This area, too, needed much more development work.

As they packed up on December 9, after their last flight

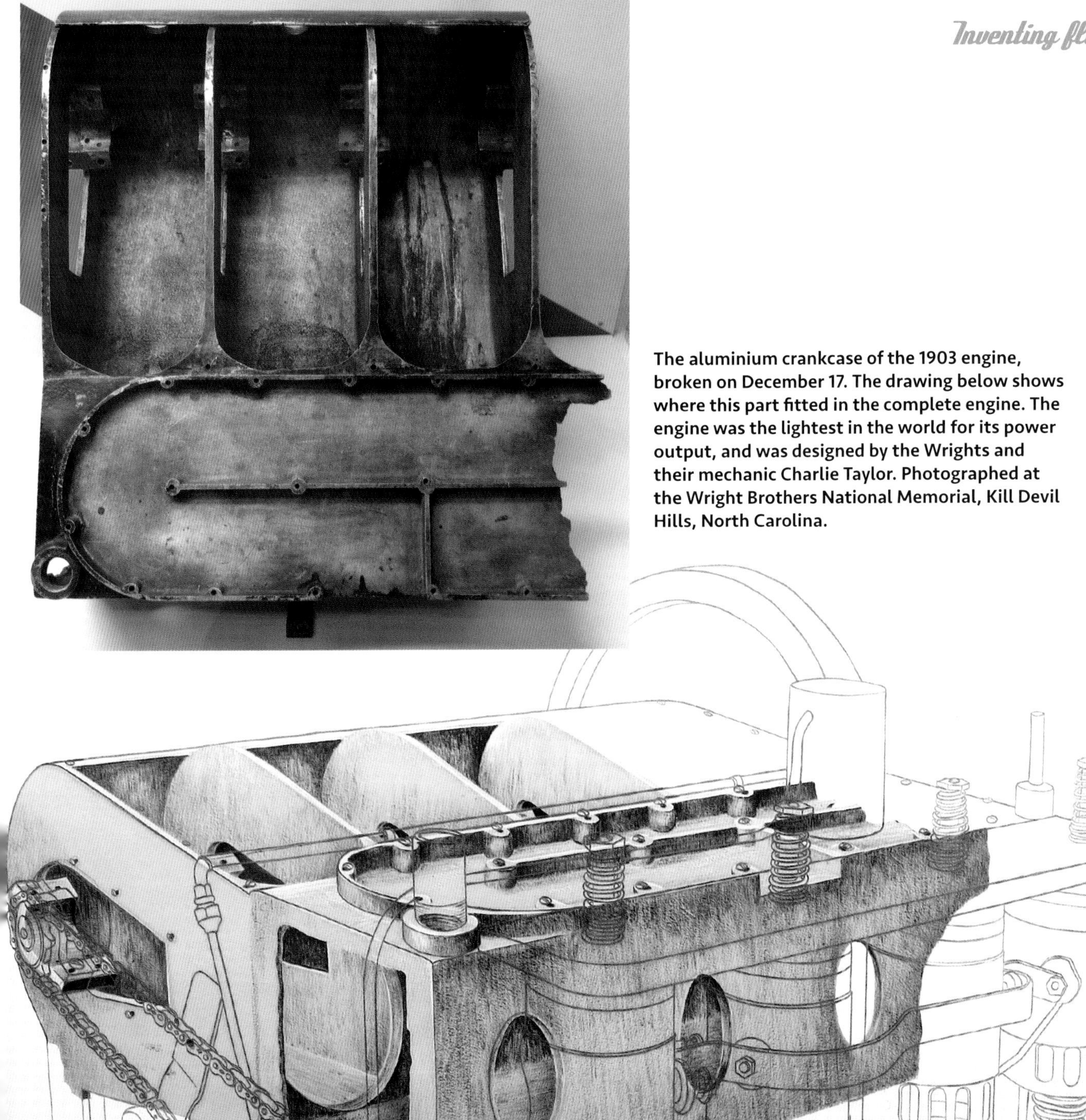

The aluminium crankcase of the 1903 engine, broken on December 17. The drawing below shows where this part fitted in the complete engine. The engine was the lightest in the world for its power output, and was designed by the Wrights and their mechanic Charlie Taylor. Photographed at the Wright Brothers National Memorial, Kill Devil Hills, North Carolina.

of 1904, the Wrights were already thinking about their next airplane.

The first practical aircraft

The Wrights began work on Wright Flyer III on May 23, 1905, and they had it ready to fly exactly one month later. Like their two previous airplanes, Wright Flyer III was a biplane with a dual canard elevator, dual pusher propellers, and a dual vertical tail. The tail was taller, and the entire craft sat slightly higher off the ground, but it was much like Wright Flyer II. In fact, the engine and almost all of the metal hardware from Wright Flyer II was reused in the all-new airframe. (They burned the old one.)

In the new plane, the brothers returned to their 1903 wing camber of 1/20 of the wing chord. Because of its now-polished cylinder walls, they found that the engine could produce more than 13kW (18hp), but catapult launches still would be necessary.

With its similarity to Wright Flyer II, the Wrights could not have been surprised when Wright Flyer III exhibited essentially the same flying characteristics as its predecessor. Flight testing began on June 23, but no flight exceeded 20 seconds, and almost every landing was hard enough to damage the plane. The worst accident occurred on July 14, when Orville was unable to control pitching oscillations. Twelve seconds after launch, he nose-dived into the ground at about 50km/h (30mph), smashing the elevators. The plane bounced three times and overturned, ejecting Orville. Dazed and bruised, he was lucky not to have suffered serious injury.

When the brothers rebuilt Wright Flyer III this time, they made some major changes. Most of their attention went into improving the elevator and getting pitch under the pilot's control. They increased the elevator surface area by over 50 per cent and, more importantly, moved the whole elevator assembly almost 1.7m (5½ft) farther in front of the wing. This moved the centre of mass farther forward, and lengthened the moment arm through which the elevator acted. The relocated centre of mass made the craft less likely to pitch upward, and the increased length made the plane less sensitive to minor elevator movements, thus reducing the likelihood that a pilot would over-correct.

After a brief first flight of the day, they reset the plane on the launching cradle, hoisted the catapult weight to the top of the tower, and cranked the engine. Wilbur took the pilot's position and released the clip holding the craft back. As the weight fell, the plane accelerated down the rail and took to the air. The spectators saw it make a graceful turn and begin to circle the field, much the same way many flights started. But instead of circling the field four or five times, Wilbur kept going.

As Amos Stauffer, one of those watching, said, "The durned thing just kept going round. I thought that it would never stop."

Wilbur flew until his fuel was exhausted; he had enough to remain aloft for 39 minutes, 24 seconds, and cover slightly over 39km (24mi), a distance longer than all the previous 109 flights put together. Once again, Wilbur and Orville Wright had succeeded in reaching their goal.

Wright Flyer III was the practical flying machine they sought. Unlike anything else in the world, it could take off, climb into the air, fly for extended periods in any direction completely under the pilot's control, and land in a safe, controlled manner. And it had shown that it could do all of this over and over again.

Six years of work, frustration, successes, failures, and dedication had led to the world's first practical airplane.

While some fundamental changes in airplane design would come in the future, the primary one being the relocation of the elevator to the tail of most planes, the basic concepts proven by the Wrights with Wright Flyer III remained the foundation for these designs.

Anatomy of Wright Flyer III

Wright Flyer III used the same basic structural design as its predecessors. The frame was primarily spruce, covered with a muslin skin. Wire cables served as diagonal members, forming a strong, yet flexible truss between the two wings. Other cables attached to the control handles, and a hip cradle transmitted the pilot's movements to warp the wings and pivot the elevators and rudders.

As rebuilt, Wright Flyer III had wings with a 12.3m (40.5ft) span; a 2m (6.5ft) chord; 1/20 camber; and a total area of 46.7m^2 (503ft^2). The wings were 1.8m (6ft) apart. The double canard elevator area totalled 7.7 square metres (83 square feet), while the dual rudders were 3.2m^2 (34.8ft^2) in area. The modified craft's overall length measured 8.5m (28ft), and it weighed 322kg (710lb).

The engine was mounted to the pilot's right, and a pair of roller chains drove the twin propellers. To counteract torque effects from the propellers, the Wrights designed them to turn in opposite directions, just as they had done on the first Wright Flyer. To accomplish this, the longer, left-hand chain went through a pair of tubes that flipped the chain between the engine and propeller.

As with the 1903 engine, the 1904-05 engine was designed by the Wrights and built by Charlie Taylor. It had several improvements over the earlier engine, including improved cooling, pressure lubrication, a fuel pump to replace the gravity feed, and compression relief for easier starting. The ingenious compression-relief device automatically moved the camshaft slightly to prevent the valves from fully closing when the cam's driving force came from the propellers, as it did when starting the engine. Once the engine started and the crankshaft began

Performance of Screws.
24 Miles pr. hr. = Speed Mach.
44 ft. sec = gross Speed.
90 lbs. = Thrust.
2 Screws 8.5 ft X 8 in.
Area of blades 5.4 sq ft.
330 Rev. Per min.
121 ft. sec = Speed of C.P.
25.3 lbs = Normal Pressure
7° = Angle of Incidence
3° = Tangential.
27° = Pitch.
Tang of 24° = 445
.445 X 90 = 40 lbs. resist. C.P.
40 X 121 = 4840 sec. lbs.
4840 ÷ 550 = 8.73 horse power.

of air under 90 lbs thrust
8½ ft = di. Screw.
112 = Area Circle, 2 screws
112 X 45 = 5000 cu. ft air.
5000 X 075 = 375 lb. air.
90 ÷ 375 = .24 lbs pr. lb.
32 X 24 = 7.68 ft per sec
45 − 7.68 = 37.32 net
37.32 X .68 = 25.3 mi per hr
= Efficiency.
90 X 24 mi. = 2160 mi. lbs.
2160 ÷ 375 = 5.76 h. p.
5.76 / 8.73 = .66 percent. Effic.
.34 = Total loss.
.20 = Loss from true slip
.14 = Loss from air moving.

As this page from their notes shows, the Wright Brothers were patient and methodical engineers. Note that just simple algebra and geometry are being used. Photographed at the Wright Brothers National Memorial, Kill Devil Hills, North Carolina.

driving the cam, it moved back to its normal position. This feature also allowed the propellers to windmill more freely in the air stream whenever the craft was gliding.

The Wrights did not consider this engine to be a particularly significant part of their achievement, but instead just an ancillary item they had to have. Others have disagreed, with Wright historian Leonard S Hobbs calling it, "the essentially perfect engineering achievement by the classic definition."

The propellers, on the other hand, were aeronautical works of art in which the brothers justly took pride. Unlike their predecessors, the Wrights recognised that a propeller was a not a fan to move air, but a rotating wing that generated a horizontal 'lifting' force called thrust. Just as they had done for their wing experiments, they used a wind tunnel to determine the shape for their propellers. The results, hand carved out of laminated spruce, were 2.6m (8.5ft) diameter propellers with an astonishingly high efficiency of 66 per cent, well beyond anything else in 1903.

For Wright Flyer III, they carved improved propellers with bent ends they called 'little jokers.' These ends looked peculiar at rest, but they flexed to a more-efficient configuration in operation. Like their earlier propellers, the brothers covered the ends in canvas to resist splitting. They utilised this design on all of their succeeding planes.

Following the 1905 flying season, the brothers disassembled Wright Flyer III and carefully packed it away. In the spring of 1908, however, they held two contracts for airplanes, one with the US Army Signal Corps, and another with La Compagnie Generale de Navigation Aérienne, a French consortium. Both contracts had performance specifications that the brothers felt they needed to test in flight, but out of the view of spectators and possible rivals. The lonely dunes of Kitty Hawk were ideal, or so they thought, so they refurbished and modified Wright Flyer III and in April took it to the site of their first flights.

The basic airframe was unchanged, but they installed one of their new 4-cylinder, vertical engines that developed about 26kW (35hp), and the pilot now sat upright in one of two seats with revised controls at his side. But in spite of Kitty Hawk's remoteness, reporters got wind of their presence and turned up to record – and exaggerate – the events, much as they had done following the 1903 flights. Nevertheless, the flights began.

On May 6, Orville took to the air for his first flight since 1905. About a week later, the brothers were surprised when Charlie Furnas, a mechanic from Dayton, appeared at the camp. Furnas wanted to fly, and they needed to test with two people on board, so Wilbur took him on a short flight on May 14, marking the first time a Wright airplane carried two people. A short time later, Furnas joined Orville on a longer flight.

That afternoon, Wilbur set off on an extended solo flight, but he mishandled the revised elevator controls take-off and nose-dived into a sand dune. He suffered only a few cuts and bruises, but the airplane was shattered. Subsequently, it was rebuilt for historic display – it can be seen at the Dayton Aviation Heritage National Historical Park.

Acknowledgement: The text of this chapter is based on material drawn from the American Society of Mechanical Engineers and is used with permission. The Wright Flyer III is a Historic Mechanical Engineering Landmark.

Chapter 3: Mutually Assured Destruction

44 years of uneasy technological peace

World War II pitched two odd alliances against one another: the Axis powers of Germany, Italy and Japan versus the Allied powers of the United Kingdom (and her Commonwealth countries), the United States and the Soviet Union. On both sides, these were marriages of necessity, but with victory achieved by the Allies, the marriage immediately started to break down. Nowhere was this clearer than in the conflict that began between the Soviet Union and her erstwhile allies, the UK and the US. The Cold War – known as such because in comparison with a 'hot war,' there was no shooting except through proxy countries – was to last from about 1947 to 1991, when the Soviet Union collapsed.

For people like me who grew up in Australia, far from the scene of potential conflict, it's hard to imagine the very real fear of nuclear war that was present in much of the world, especially in the 1950s and 1960s.

Probably the biggest awakener for me was Nevil Shute's 1957 novel *On the Beach*, a book set in Australia. In the novel, the northern hemisphere countries are destroyed by nuclear war, with the last vestiges of humanity surviving far from the conflict in the southern hemisphere. But a rolling tide of nuclear fallout is moving southwards, wiping out all living humans before it. The novel – a masterpiece of terror set in the midst of suburban humdrum – features a US submariner who is sent northwards to explore the extent of devastation. He finds cities that look completely unmarked, even with neon signs still glowing via the power of hydroelectric stations – but everyone is dead through radiation poisoning. Another novel I read as a teenager – *Fail-Safe* – explores what would happen if a nuclear attack was accidentally launched – a distinct possibly both then and now.

At the time these books were written, and for decades afterwards, the nightmare scenarios were absolutely

Nevil Shute's 1957 novel *On the Beach* alerted many people – including me – to the consequences of a widespread atomic war. It's a novel that still has tremendous power. (Owned by the author)

possible … especially given the incredible nuclear arms race that then developed between the US and Soviet Union. With each country easily able to destroy the other, the acronym 'MAD' was developed – Mutually Assured Destruction. In this doctrine, neither party will launch a

Part of the social ephemera on display at the National Atomic Testing Museum in Las Vegas. Note the September 1964 comic – *The Atomic Superboy*.

nuclear attack because it will be certain that, in turn, its country will also be destroyed. However, for MAD to be effective, the first country must either be able to detect with absolute certainty the launch of a nuclear attack, or be able to resist the first stage of a nuclear attack in order the retaliate. (You can see why the moniker 'MAD' resonated!)

Initially, the atomic bomb deliveries were to be made via long-range bombers, then Inter Continental Ballistic missiles (ICBMs) were developed. Atomic missiles able to be carried by nuclear submarines were then built; as these submarines were virtually undetectable and could stay underwater for very long periods, their deterrence value was high.

In the period since the Cold War finished, some amazing museums have been created that show the technology of this era. The best that I have been to is a decommissioned Titan II missile silo near Tucson, Arizona, USA. It's one of 54 Titan II missile sites that were on alert across the United States from 1963 to 1987. Another extraordinary site is the Cold War Bunker in York, United Kingdom. With the UK so much geographically closer to the Soviet Union than the US, authorities set up protected, underground monitoring stations from which a nuclear attack on the UK could be monitored. It's incredibly sobering to realise the extent of preparations for what many people saw as an inevitable war.

More subtle is the social coverage of nuclear war (and all things atomic) that occurred in the 1950s and 1960s, perhaps so as to allay what otherwise would be social unrest as countries like the US headed towards an abyss. In the National Atomic Testing Museum in Las Vegas, displays of real nuclear bombs are sobering, but for me most fascinating was the contemporary social ephemera. From children's Atomic Energy Labs to 'how to' guides on avoiding radiation during an atomic bomb attack; from cereal packets promoting competitions to win an Atomic Bomb Ring, to recipe books for atomic cocktails. The overall effect was to suggest a widespread societal normalisation of atomic bombs – and their use.

I am not a pacifist, but MAD seemed hell-bent on killing the patients in order to save them from the disease …

YORK COLD WAR NUCLEAR BUNKER

Designed to monitor the devastation as nuclear bombs rained down on the United Kingdom

The city of York in the UK is a small city – population 140,000 – with an old past. Founded by the Romans in AD71, by the Middle Ages it had become a major wool trading centre. Fast forward to the 19th century, and it was the hub of the railway network. You can still see evidence of that past – the defensive wall around the city, and the fantastic railway museum.

But what the potted histories often don't mention is that York was also home to one of a string of underground bunkers built in the 1950s across Britain. The bunkers were designed to monitor the nuclear attack that may have emanated from the hostile Soviet Union, located just a (relative) hop and a skip away. These are places from which it was fully expected that observation and monitoring would be made of a dying country – one enveloped in poisonous and spreading radiation, killing all those in its path.

The deployable communications antenna was erected only when communication was necessary. Keeping it retracted at other times prevented it from blast and electromagnet pulse damage.

The York Cold War Nuclear Bunker is positioned at the end of what is now a quiet suburban street. However, when it was built, it was located in an orchard. Only the entrance is above ground. A very good tour is available of the facility.

So let's go inside the York Nuclear Bunker, a place redolent of circumstance almost beyond imagination.

Despite it now being in the midst of houses, when the bunker was built, it was on the site of a former orchard of a mansion, in a hollow hidden from a nearby busy road. The bunker – one of a type officially called a 'semi-sunken Group Headquarters' – used a 'cut and cover' construction technique. This involved digging a pit, building the concrete structure and then covering most of it with earth. At least 1m (3ft) of earth was placed over the concrete – this was designed to increase protection against bomb blast and heat, and to provide some defence against radiation.

The entrance block is the only visible part of the three-level structure. It was originally painted white to reflect the heat from nuclear flashes. The door, designed to withstand blasts, opens to a passageway divided by an air lock. The air lock used two rubber-sealed, gas-proof doors designed to prevent nuclear radiation particles from entering. On the right is a telescopic radio antenna that could be raised when signals needed to be sent and received. The antenna is surrounded by an earthed cage to protect against electromagnetic impulse damage. At the opposite end of the entrance block was an emergency escape hatch that used a counter-weighted steel manhole cover. The hatch was designed to withstand a compressive force of 30t.

The bunker had three permanent staff, but during times of emergency, this grew to 50-60 people. Only a small kitchen was provided. Note the air-conditioning vent – the ventilation system could be operated in a special filtration mode to catch radioactive particles.

Filtered air conditioning system

The bunker used extensive air-conditioning and ventilation equipment. Ventilation could be configured in four modes: 'normal' (fresh air mixed, cooled and circulated); 'modified normal' (air-cooling system turned off to save water); 'recirculation' (internal air re-used); 'filtration' (air drawn in through particulate filters to remove dust and radioactive particles). Air conditioning was important as the outside atmosphere may have been heated by the nuclear blasts. Power came from a diesel generator – the tank had enough fuel for 40 days of continuous use. Air compressors were used to power the waste ejection system – because the bunker was located below the level of the sewer, waste needed to be periodically blasted upwards from holding tanks.

During its life (it was opened in 1961 and closed in 1992) the York bunker was staffed by three permanent staff and, during times of emergency, 50-60 Royal Observer Corps volunteers, scientists and telecommunications engineers. These worked in three shifts – duty, standby and rest. Little sleeping accommodation was provided. A dormitory comprised six bunks while another had provision for four people, necessitating 'hot bedding' – people used whichever bed was available. During training, the bunker was expected to operate in a 'nuclear environment' for at least two weeks. A small kitchen and canteen were also provided.

The bunker had three main functions: plotting where atomic bomb bursts had occurred; assessing their magnitude; and then plotting the extent and movement of a radioactive fallout cloud. Charts were used to plot the location and size of the bomb bursts.

The bunker was equipped with instrumentation to allow monitoring of the outside environment. One such instrument was a Bomb Power Indicator – this comprised a simple pressure gauge. On the surface, the indicator used a pair of circular steel plates attached to a galvanised steel pipe that ran to the indicator. The indicator, reading up to 5psi, using expanding metal bellows to drive the needle. The needle remained at its maximum until reset – a process that would normally occur after 10 seconds in order to record further explosions.

Ground Zero Indicator

A Ground Zero Indicator was also located on the surface. It comprised a cylindrical pinhole camera using four 5mm holes, with one hole each facing north, south, east

This bomb power indicator was connected to the surface and measured over-pressure in pounds per square inch. A peak-hold device, it was reset to await the next nuclear bomb explosion.

The ground zero indicator was mounted on the surface. It comprised a cylindrical pinhole camera with holes facing north, south, east and west. A cylindrical piece of photographic paper recorded the location of the nuclear explosions. The paper needed to be physically retrieved, necessitating exposure of the retriever to lethal radiation.

and west. Inside the cylinder the space was divided into four semi-circular quadrants, each equipped with a piece of low sensitivity photographic paper. The paper was mounted with a plastic sleeve on which was marked a grid indicating bearings and elevation. If a nuclear explosion

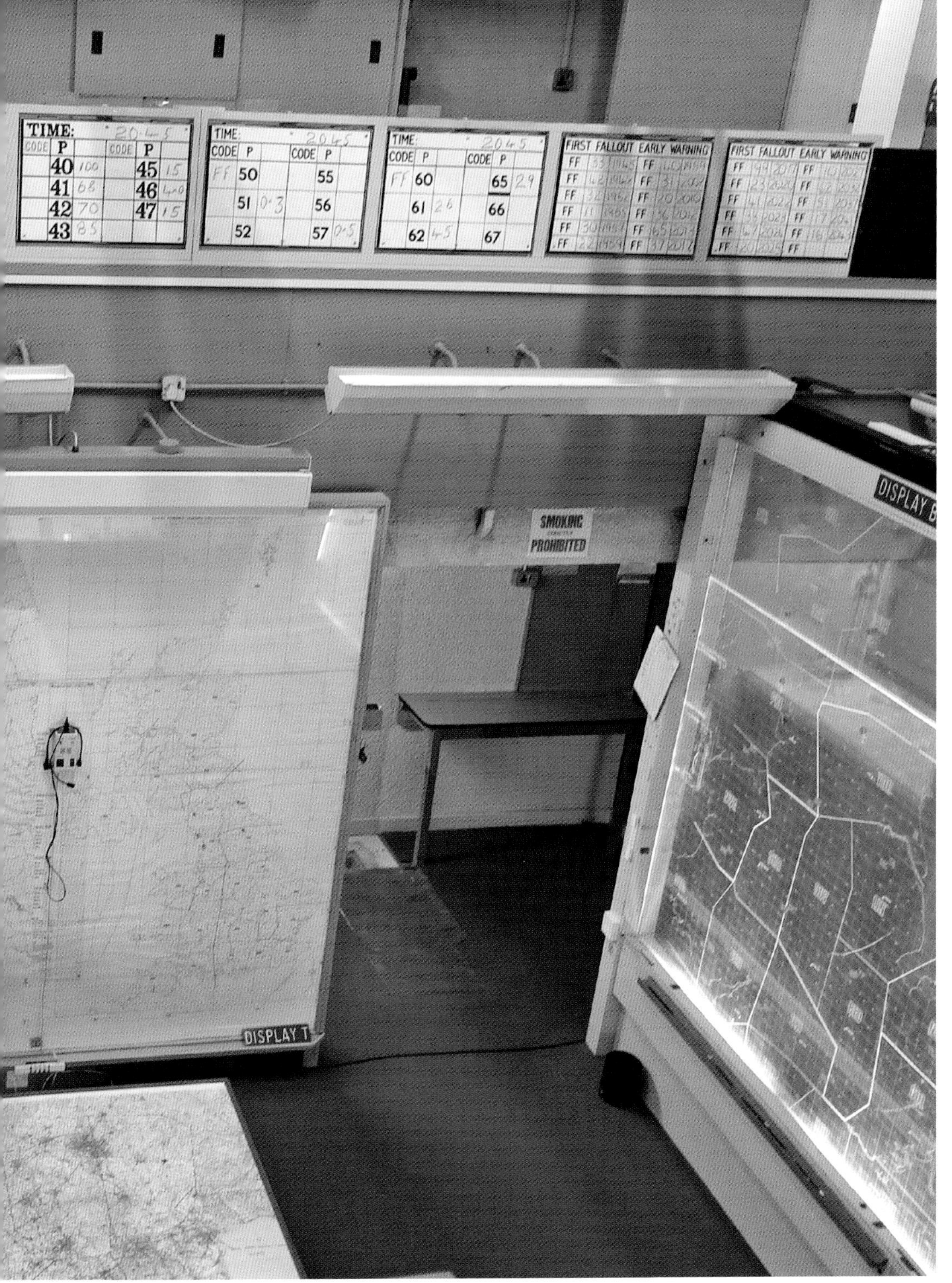

The main operations room within the bunker. The situation was plotted on large maps using chinagraph pencils. Reports were telephoned in from smaller three-person underground monitoring posts built across the UK.

occurred, a spot would be burnt in the paper. By the use of several of these devices located at different reporting posts, the Group Headquarters could determine the location of the bomb. Note that following a bomb blast, the photographic paper had to be physically retrieved from the device: this would have exposed the retriever to potentially lethal radiation. Only rudimentary decontamination facilities were provided at the bunker.

These extraordinarily primitive detectors were later supplemented by an electronic detector. The 'Atomic Weapon Detection Recognition and Estimation of Yield' (AWDREY) device was designed to detect both the optical and electromagnetic signatures of a nuclear explosion, and set off automatic alarms, including transmitting data about the explosion to other group and sector headquarters. Twelve units provided coverage of the whole of the UK. From 1974, a further electronic detection system gave enhanced direction capability.

Within the bunker was a main operations room. At the map desk sat the duty controller and senior warning officers. Behind the desk were two large maps printed on transparent plastic. Display A showed the current situation; Display B showed the cumulative situation, plotted at two-hour intervals. Both maps were for the local area. In addition,

(continues on page 40)

The dropping of the atomic bomb – nicknamed 'Little Boy' – on Hiroshima on August 6, 1945, set the scene for the Mutually Assured Destruction doctrine of the ensuing Cold War.

The atomic bombs (the alternative design, 'Fat Man,' was developed simultaneously) were the culmination of the greatest scientific and industrial effort ever made. Once the scientific basis for a potential super-weapon had been established, the US threw simply staggering resources at the development of the atomic bombs. Even with the retrospect of history, it seems doubtful that any other country could have marshalled sufficient resources to develop a working weapon in the time in which the US achieved that outcome.

Three complete cities – Los Alamos, Hanford and Oak Ridge – were built from scratch, and these were just some of the locations where 130,000 people worked to build the bomb. The scale of the effort, and the huge number of unknowns regarding the technology, meant that factories were built for techniques that had not yet even been developed. Information was strictly compartmentalised, and so many operators didn't even know what the machines they were controlling were actually doing. Nothing was out of reach: when there was insufficient copper available to build the electromagnets needed to separate isotopes, the US treasury supplied 14,700 tons of silver bullion to wind the coils. In recent values, the cost of producing a working atomic bomb is estimated to have been something in the order of US$22 billion.

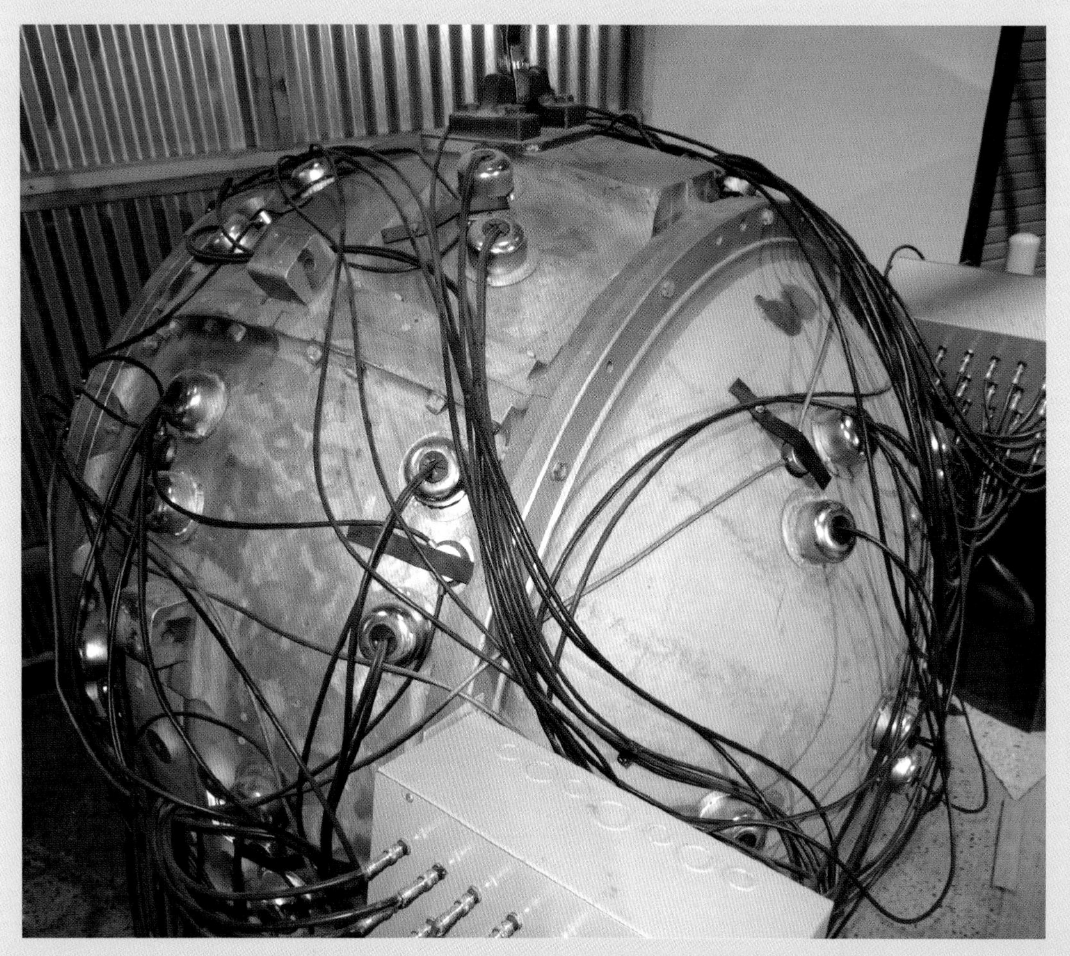

A mock-up of the world's first atomic device, nicknamed Gadget. It was called a 'device' because it was not yet a deployable weapon and words like 'atomic' or 'bomb' were avoided for security reasons. Gadget was successfully exploded at the Trinity test site near Alamogordo, New Mexico. It was an implosion-type plutonium bomb, similar in design to Fat Man used three weeks later in the attack on Hiroshima. Photographed at the National Museum of Nuclear Science & History, Albuquerque, New Mexico.

Desert sand scooped up into the fireball created by the Trinity atomic test melted inside the mushroom cloud and then fell back to earth. There it solidified into puddles of a new material, born by atomic blast – Trinitite. On display at the Bradbury Science Museum, Los Alamos.

Enola Gay, the Boeing B-29 Superfortress bomber that dropped the atom bomb on Hiroshima. It is on display at the Steven F. Udvar-Hazy Museum in Virginia, USA.

Developed in the 1950s, the M65 atomic cannon – better known as Atomic Annie – was a towed artillery piece capable of firing a nuclear weapon. The gun and carriage weighed 83t and used two tractors, one at each end, that were capable of being independently steered. Only one projectile was ever fired from the cannon, a 15 kiloton warhead that detonated at a distance of 11km (7mi). Photographed at the National Museum of Nuclear Science & History, Albuquerque, New Mexico.

Significantly closer to the old USSR than most cities is Gothenburg in Sweden, where you'll find the Aeromuseum. This comprises an enormous underground series of tunnels, used primarily during the Cold War to house a Swedish air force squadron. The underground base was decommissioned only about 15 years ago, and so is in as-new condition. The ventilation system, the curtains, the steel blast doors – all designed to protect from nuclear attack – appear to be in perfect working order. Even the doors to what are now the public toilets are steel and are skirted with rubber to provide a seal against airborne contaminants. It's well worth a visit to see just the bunker – the museum's contents are a bonus.

Entering service in 1948, the Convair B-36 was a primary nuclear weapons carrier for the US Air Force. It was the largest mass-produced piston engine aircraft ever made. Six 28-cylinder Pratt & Whitney R-4360 Wasp Major radial engines were mounted in an unusual pusher configuration, and later a pair of General Electric J47-19 jet engines were suspended outboard of the propellers. The aircraft was in operational use from 1949 – 1959. This one is pictured at the Pima Air & Space Museum, Tucson, Arizona, USA.

The extraordinary investment in defence by the US (and of course the USSR and their respective allies) is best illustrated by a tour of the 'boneyard' at Davis-Monthan Air Force base in Tucson, Arizona. Here, US defence aircraft are stored for potential later use, or prior to being scrapped. Typically, there are more than 4400 aircraft in storage at any one time. Bus tours of the site are available – and are recommended.

there were maps for Europe and the United Kingdom. The clear plastic maps were manually updated from behind, using chinagraph pencils.

The information was relayed to the plotters and warning officers by post display operators receiving information from outlying underground monitoring posts. (The underground monitoring posts were three-person units equipped with pressure and photographic paper indicators. Hundreds of these underground posts were built across the UK.) The details were written by operators on the white boards; the boards were rotated every five minutes so that they could be read by those working within the 'well' of the room.

A sound-proof communication room was located adjacent to the 'well.' Equipped with machines such as teleprinters, the equipment was designed to have redundancy – coping with the failure of certain communication channels. Information was transmitted to other Group Headquarters and the Midland Sector Control. In the late 1980s, the telecommunications equipment was upgraded and the manual display boards replaced by desk-top computers and monitors.

The whiteboard gives a good feel for the operational situations likely to be met. It includes sections to indicate the status of standby power, ventilation and mains water; whether the shelter is being subjected to nuclear fallout; the remaining number of ration packs; and stocks of fresh water and diesel.

The York Nuclear Bunker is an extraordinary indicator of the temperature at which the Cold War operated. One of 29 such facilities constructed across the UK, the York bunker was part of observation and monitoring system that in itself, was just part of a broader military and civil defence response to the Soviet threat.

TITAN II MISSILE SILO

A massive underground complex designed to withstand nuclear attack and launch a devastating retaliatory strike

The whole doctrine of Mutually Assured Destruction required the ability to strike back after sustaining a nuclear attack. To do that, you needed retaliatory mechanisms that could withstand a targeted nuclear assault – and still function. The Titan II network of missile silos provided part of that assurance. Associated with air force bases in Arkansas, Kansas and Arizona, USA, the missiles were sited in underground firing complexes, located far enough apart that a direct nuclear attack on one silo was survivable by the others. There were 54 Titan II missiles, with 18 in each state.

Enormous underground complexes

The missile complexes were enormous. Each firing complex consisted of three underground parts – a silo that housed the vertically-standing missile completely underground, a control room, and a central blast lock and access portal containing lifts (elevators), decontamination showers and crew rest areas. Underground corridors, called cableways, connected the different parts.

Construction of each site started with scooping out a 'bathtub', some 120m (440ft) long, 60m (200ft) wide and 12m (40ft) deep. The silo, control centre, blast lock and access portal were all contained in this excavation. The silo was excavated at one end of the bathtub, going down a further 33m (110ft) for a finished depth of about 46m (150ft). This excavation was made by a bulldozer at the base of the silo working in conjunction with a lift-bucket – a process known as 'scrape and lift.'

The silo wall was 1.2m (4ft) thick for much of its depth, with the top 9m (30ft) increased in thickness to 2.5m (8ft). At the base of the silo, a huge wedge-shaped concrete blast deflector was constructed. Two large exhaust vents carried exhaust gas from the missile's engine to the surface. The launch duct was lined with special sound-absorbing modules. These, and a water spray system, helped protect the missile from damage caused by high-intensity sound and vibration. The missile sat on a ring-like thrust mount, supported on four enormous springs. These allowed the missile to withstand the shock wave of nuclear attack.

760 tonne door

A large #-shaped reinforced concrete beam sat on top of the silo, helping to support the silo door. The door, weighing 760t, had a top surface 75mm (3in) thick. It was moved by hydraulic jacks, and could be fully opened in about 20 seconds. Much of the complex's construction was in steel-reinforced concrete, with more than 6400 cubic metres (8400 cubic yards) used at each site. More than 2200t of steel rebar was employed, with much of it an incredible 60mm (2.25in) thick.

Titan II was the largest and most powerful Intercontinental Ballistic Missile ever built by the US. The missile had a huge 9-megaton nuclear warhead (ie 9000 kilotons, or 600 times the size of the Hiroshima bomb) that could be delivered to targets more than 10,000km (6000mi) away in about 35 minutes. Each missile was capable of devastating an area of about 2300km^2 (900mi^2). The United States' nuclear doctrine planned on using the Titan IIs as a deterrent that targeted large Soviet military bases and civilian population centres. The missile's top speed was 25,700km/h (16,000mph) – more than 20 times the speed of sound.

Launchable within a minute

Unlike earlier US missiles, the Titan II did not use liquid oxygen. Liquid oxygen boils away rapidly at normal

It looks toy-like, but the Titan II missile pictured on the right in its underground silo is 31m (103ft) tall. 54 of these enormous underground facilities were built in the US, each housing one missile. (Courtesy Titan Missile Museum)

temperatures, meaning that it cannot be stored in the missile, and instead must be loaded aboard the missile just before launch. Loading liquid oxygen took precious minutes that delayed retaliation and made the missiles vulnerable to attack. Instead of liquid oxygen, Titan II used two chemicals that were stable at temperatures of around 15.5°C (60°F). This allowed Titan II to be stored in its air-conditioned silo, fully-loaded at all times, and ready to launch in less than a minute. In addition, Titan II was the first US missile that could be launched directly from its silo, further reducing vulnerability to attack.

Titan II was 31m (103ft) tall and 3m (10ft) in diameter. The re-entry vehicle, containing the atomic weapon, sat on top of the missile (it is often referred to as the nose cone). A two-stage rocket, Titan II was fuelled with a 50-50 blend of hydrazine and unsymmetrical dimethyl hydrazine, having the brand name of Aerozine 50. The oxidiser was nitrogen tetroxide. The two dangerous and toxic chemicals had to be kept separate at all times until launch. Inside the thrust chambers, the fuel and oxidiser ignited spontaneously in a process called hypergolic action – no spark or external ignition source was required.

The Titan II in its silo at the Titan Missile Museum, Arizona. The rectangular cut-out in the re-entry vehicle is to demonstrate to nuclear weapons inspectors that this is a deactivated missile.

The Stage 1 engine developed 20,000kg (430,000lb) of thrust and burned until it ran out of propellant after about 2.5 minutes. It was then jettisoned as Stage 2 ignited. The Stage 1 engine consumed about 96,000 litres (25,000 gallons) of propellant (640 litres or 170 gallons per second) and boosted the missile to an altitude of about 80km (50mi). Stage II burned for about 3 minutes and propelled the missile to an altitude of about 300km (200mi). It was then shut down at the appropriate time by the on-board guidance system. When the Stage II engine shut down, two small solid-fuel vernier motors fired to fine-tune the speed and direction of the missile. Stage II then separated from the missile and the re-entry vehicle, carrying the nuclear weapon, continued towards the target in a pre-planned ballistic trajectory. The weapon was the only part of the missile to reach the target.

A Titan Missile Museum guide seated in front of one of the massive set of nested springs that support the underground control centre. These springs allowed the complex to withstand the earthquake-like vibrations of nearby nuclear attack. The tour is excellent.

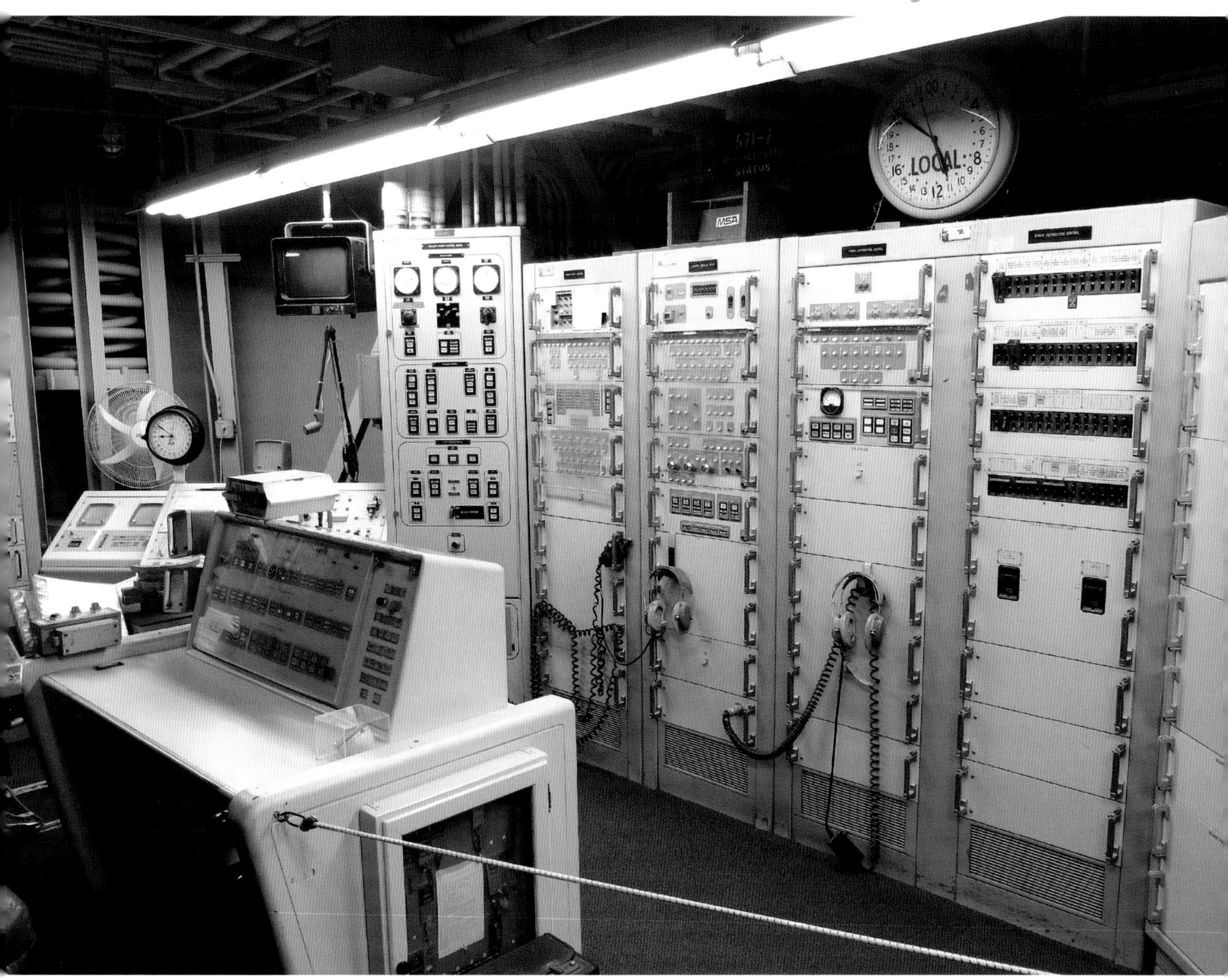

The control room from which Armageddon could be launched. Each Titan II missile had a huge 9-megaton nuclear warhead (ie 9000 kilotons, or 600 times the size of the Hiroshima bomb) that could be delivered to targets more than 10,000km (6000mi) away in about 35 minutes.

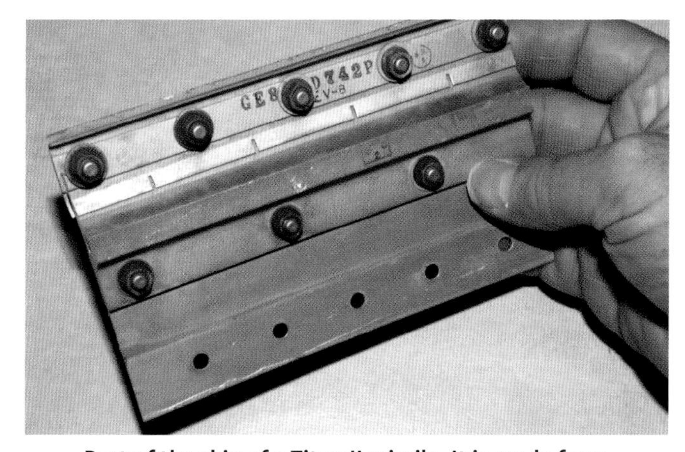

Part of the skin of a Titan II missile. It is made from 2014-T6 aluminium, an alloy of aluminium and copper. (Owned by the author)

Titan II entered active service with the US Air Force in 1963. Following decommissioning in the mid-1980s, the remaining missiles were converted into space-launch vehicles for satellites and were used in that role until the final Titan II was launched in 2003.

The Titan Missile Museum in Arizona comprises a complete launch complex. Excellent quality guided tours are available to take participants through most of the underground facilities. The missile museum is one of the most extraordinary places to visit in the world.

POLARIS LAUNCH DESK – Launching mass death with the press of a button

As with much of the military technology of the era, fire control system for the Polaris missiles appears surprisingly primitive. With its analog meters, chunky knobs and filament pilot lamps, it looks more like the control panel of a small power station than the submarine control desk from which tens of millions of people could be obliterated.

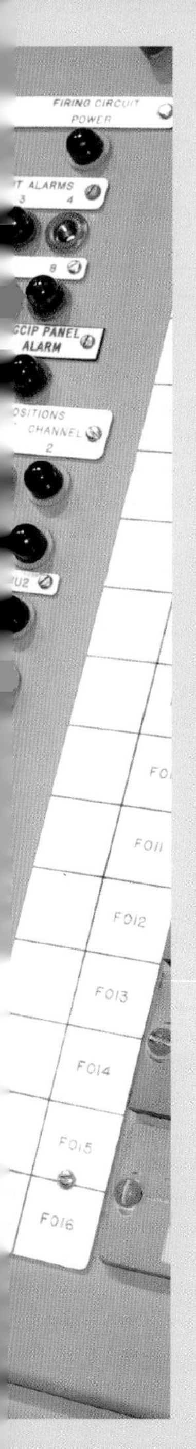

The Polaris was a submarine-launched ballistic missile designed in the late 1950s, in service up until 1996. A solid-fuel design, it carried three nuclear warheads giving a 600kt yield for each missile. (The atomic weapon exploded over Hiroshima had a 15kt yield.) Each submarine equipped with Polaris missiles carried 16 missiles; enough to potentially devastate an entire country.

The USS Bowfin Submarine Museum in Honolulu, Hawaii, has a display of the Mark 80 fire control system that was used to monitor, arm and fire Polaris missiles.

One of the most sobering knobs you will ever see. Missiles 1 through to 16 can be selected for firing, or the knob can be set to 'all.' The latter would have been selected only in the event of an all-out atomic war where literally hundreds of millions would die.

Chapter 4:

Turbine power

Round and round, not up and down

Turbines: for most people, the term conjures up little. Even when we add the dictionary definition ("a rotary mechanical device that extracts energy from a fluid flow and converts it into useful work") many will still struggle to see much of interest.

But turbines power nearly every aspect of how we live. From steam, gas and hydro turbines used in power stations, to the jet engines that propel aircraft, to steam turbines in ships – turbines are one of the main prime movers that make our technological world work. And of course, many turbines operate in conjunction with other designs of engine – the turbochargers on petrol (gasoline) and diesel engines are turbines, too.

The steam turbine was invented by Charles Parsons in the 1880s. Fascinatingly, rather than demonstrate his new design of engine with something as boring as a dynamometer test, he installed it in a purpose-built, small ship called Turbinia. It was soon the fastest ship in the world.

Turbinia is in the Discovery Museum in Newcastle upon Tyne, UK. It was in the late afternoon winter darkness when we found the location. Suffering a travel bug, my wife Georgina and son Alexander initially stayed in the car, so I went into the museum alone. When I saw Turbinia, still looking rakish and fast nearly 125 years later, tears welled in my eyes. What a ship! What an engine! What bravery and brilliance! Then, when I saw displayed nearby one of the rubber-band powered models that Parsons had tested to find the hydrodynamically slipperiest hull shape, my admiration increased still further. What elegant engineering!

Elegant, but in a quite different way, are the enormous hydro turbines in the power station at the Hoover Dam on the Colorado River, on the border between the US states of Nevada and Arizona. Built in the 1930s Great Depression, you'd think the power station would be 'low-budget utilitarian.' But not a bit of it! Instead, the huge generator hall is superbly finished with a terrazzo floor, with the exposed parts of the turbine housings painted a rich dark red. Art deco touches are everywhere on the dam and power station. (And, while it's not usually stated, one can see strong similarities to 1930s German Nazi architecture, especially in the Winged Figures of the Republic statues.)

Tiny in comparison with the Hoover Dam turbines are those turbines fitted to cars. In the 1960s, car makers saw gas turbines as the motive way of the future. Rover in the UK developed several prototype cars – including a Le Mans racer – while Chrysler in the US went a step further and actually built a production run of turbine-powered machines. These cars were lent to members of the public for evaluation – incredibly, turbine powered cars were being driven on public roads.

But for most people, the closest they come to turbines is when travelling in a jet-powered plane. Invented in the mid-late 1930s by Frank Whittle in the UK and Hans von Ohain in Germany, development and adoption of the radical new jet powerplant was slow. It was especially slow when considered in retrospect – just a few years after World War II, the adoption of jets for larger aircraft (both civil and military) was near-universal. To look at those first jet engines is to see an extraordinary mix of refined engineering and machining (for example, the huge centrifugal compressors) and, in contrast, what looks rather primitive (the sheet metal burners located around the engines' peripheries). Within a short period, the centrifugal jet designs were gone, replaced by axial compressors and turbines that haven't changed enormously to this day.

From the tiniest car turbo that you can hold in your hand to the enormous turbines found in power stations, they're all machines that reflect the heritage of Parsons, Whittle and Ohain.

Turbinia – in 1894 the fastest vessel afloat – used a steam turbine engine, the first use of such an engine. It stunned the world, none more so than the British Navy that had no vessels anywhere near as fast. Photographed in the in the Discovery Museum in Newcastle upon Tyne, UK.

GENESIS OF THE JETS

Two inventors – and they were on opposing sides in World War II

Frank Whittle, an officer in the Royal Air Force (RAF), took out a patent for a jet engine in 1930. Reporting his patent to the government's Air Ministry, as he was required to do, Whittle found little interest in his radical new design. As a result, the invention was not classified and details soon began appearing in technical journals, including those in Germany and the US.

Whittle applied for, and was accepted, to undertake engineering studies at Cambridge. During this time, his patent came up for renewal, but Whittle couldn't afford the £5 fee. However, two ex-RAF officers came to the rescue, and together with other investment, the firm Power Jets Ltd was formed in 1936.

Whittle and his pioneers had recurring problems in the combustion and turbine sections of their engines. These included turbine failures and engine explosions. Finally, in 1939, the Air Ministry awarded Power Jets a contract for an experimental jet engine to power a specially designed aircraft – the Gloster E28/39. The engine, designed W.1, was to develop 400kg (885lb) of thrust.

Meanwhile, in Germany, a jet aircraft was about to fly. Hans von Ohain had been investigating the possibilities of the jet engine for aircraft, and in 1935, he had the first experimental jet engine built in a car repair shop. In

April 1936, Ohain joined the aircraft company Heinkel. Ernst Heinkel, who was particularly interested in fast aircraft, recognised the possibilities of jet propulsion. Test runs of the HeS 1 hydrogen-powered demonstration engine showed such positive results that in 1937 the construction of the first aircraft jet engine was started. This engine was the HeS 3. The HeS 3, producing 500kg (1100lb) of thrust, was then installed in the purpose-built He 178 test aircraft, with a successful first flight taking place on 27 August 1939 – the first by a jet-powered aircraft.

Meanwhile, in Britain the Gloster E28/39 had its maiden flight on 15 May 1941. The Gloster Meteor, Britain's first operational jet fighter, entered service with the RAF in July 1944. The Messerschmitt Me 262, Germany's first jet fighter, had entered service three months earlier, in April 1944.

A Power Jets W.1 jet engine. The W.1 was built under contract by British Thomson-Houston in the early 1940s. Photographed at the London Science Museum.

A replica of the He S 3 B engine, constructed by Hans von Ohain. The original was built in 1939 but was not preserved. The engine has a mass of 360kg (800lb), a diameter of 1.05m (41in) and a thrust of 43kg (95lb). It was not designed to fly but to instead prove the concept of a jet engine. Photographed at the Deutsches Museum, Munich, Germany.

Rolls-Royce Nene jet engine cutaway. This engine was first run in late 1944 and had a thrust of about 2250kg (5000lb), making it then the most powerful jet in the world. Here you can see, near the middle, the very large, centrifugal compressor used in these early jets, making their package 'short and fat' rather than 'long and skinny' like later axial designs. This engine was fitted to the De Havilland Vampire and Hawker Sea Hawk, among other aircraft. It was also sold to the Soviet Union – when the USSR was still an ally of Britain – and then, when the Cold War began, was reverse-engineered into a modified and enlarged version that powered the famous MiG-15 fighter. This engine was photographed at the Fleet Air Arm Museum, Somerset, UK.

My, how you've grown! A GE90 engine. It is 3.4m (135in) in diameter, 7.3m (287in) long and has a thrust at sea level of 53,000kg (115,300lb). This engine is fitted to the Boeing 777 and is among the most powerful of jet engines available. Photographed at the Future of Flight Aviation Center, Seattle, USA.

Prior to the jet engine being developed, turbines were flying – as turbosuperchargers. Turbosuperchargers – today we'd simply refer to them as 'turbos' – were used to force-feed piston engines, allowing them to fly at high altitudes without power loss. That is, the forced aspiration compensated for the thinner air.

General Electric (GE) in the US were on the leading edge of this technology. Their first turbosuperchargers were developed in 1916 for aircraft being used by the US Army. Test specifications called for the supercharged engine to be run at full power at sea level for 15 minutes, but GE engineers had a better idea. If the turbosupercharger was to work at high altitudes, why not test it in those conditions? The highest accessible mountain in the US, Pikes Peak, was chosen as the test site. The Liberty engine, that developed 260kW (350hp) at sea level, was found to develop only 172kW (230hp) at the 4300m (14,100ft) altitude of Pikes Peak. But when the turbosupercharger was activated, the engine developed 265kW (356hp) – a little more than achieved at sea level.

GE continued development of the turbosupercharger through the 1920s, proving its advantages over direct-driven, geared superchargers. By the 1930s, as aircraft flew higher, there was renewed interest in turbosuperchargers for both passenger and military applications. The Boeing B-17 Flying Fortress was the first aircraft equipped with the GE turbosuperchargers, allowing the engines to develop full power to well above 7600m (25,000ft). Subsequently, the turbosuperchargers were fitted to aircraft such as the P-38 Lightning, B-24 Liberator, P-47 Thunderbolt and then later to the B-29 Superfortress.

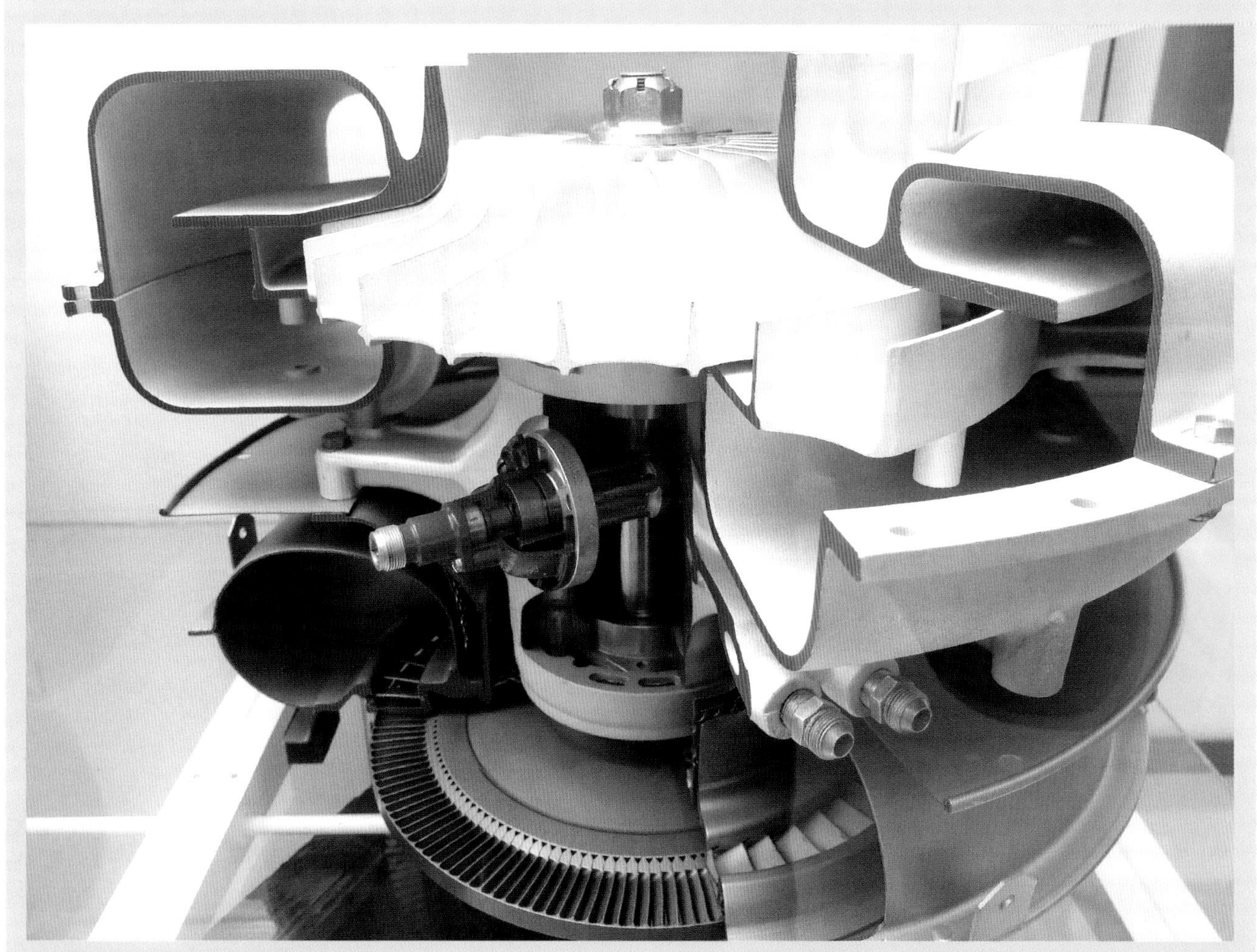

Cutaway view of a General Electric Type C Turbosupercharger, as used in World War II aircraft. It was photographed at the Steven F. Udvar-Hazy Museum, Virginia, USA.

Incidentally, if you're wondering at the long periods between the first flights of jet-powered aircraft in both Britain and Germany, and the time that operational aircraft were available, it was largely because the authorities in both countries dragged their heels. In Germany – the absolute hotbed of engineering development in World War II – Hitler had been so confident of winning the war that he stopped for a time all long-term engineering development of new technologies. And in the UK? The book *Jet – the Story of a Pioneer* by Sir Frank Whittle, covers in detail how Whittle strived to gain finance, fought government backstabbers, overcame ignorance – and tried to stop other companies that were making parts for the engines inventing 'improvements' (that weren't)!

THE FIRST STEAM TURBINE – Starting a revolution in electricity generation

Charles Parson invented the steam turbine in 1884. Parsons was a son of the Earl of Rosse, an eminent Victorian astronomer, and learned much engineering and science from him. He studied mathematics at Cambridge, but, unusually for the time, then entered Sir William Armstrong's works as a premium apprentice to add to his engineering education.

In the turbine engine, the expanding steam works continuously over many turbine stages as it passes from one to the next. This gives far higher efficiency than was then possible with the steam piston engine. The steam turbine also combines freedom from vibration with the high speed suited to electrical generators.

Parsons also pioneered the use of steam turbines in ships. To promote this, he gate-crashed the Naval Review in 1897 at Spithead with his vessel Turbinia, steaming through the lines of naval ships at 30 knots (55km/h) and outpacing patrolling naval vessels. For some 50 years, until the advent of the large diesel, the steam turbine was the standard powerplant for large vessels.

Steam turbines are used in all coal, gas and nuclear power stations, producing about 80 per cent of the world's electricity.

This is the first experimental steam turbine prototype produced by Charles Parsons in 1884. It's only about a metre long, and was photographed in the Science Museum, London, UK.

HOOVER DAM'S MIGHTY HYDRO-ELECTRIC TURBINES

Engineering elegance at its best

The Hoover Dam is about 50km (30mi) from Las Vegas, on the Arizona/Nevada border, USA. It's one of the most impressive dams in the world, and was completed in 1936. It powers the turbines of a large hydroelectric power station – but first, the dam wall itself.

Measured from foundation rock to the road on the crest of the dam, the wall is 221m (726ft) high. At its base it is 201 metres (660ft) thick, tapering in thickness to 'only' 14m (4ft) at the top of the wall. The dam wall, constructed of concrete, weighs over 6 million tons. During the concrete pour, which was conducted in blocks, pipes were embedded in the concrete to allow it to be cooled during curing. Over 21,000 workers toiled to build the dam; about 100 of them died during the dam's construction.

The dam's purpose is to hold back the Colorado River, so forming Lake Mead, the largest man-made lake in the US. Lake Mead has a surface area of 146,000 acres (59,000 hectares) and has a length of 185km (115mi). The reservoir can store up 2 years average water flow of the Colorado River. At the base of the wall, the water pressure is 21.5bar (312psi).

Integral with the design and construction of the dam wall are its two huge power stations. These are in the form of a U-shape structure located at the base of the dam. Each power station wing is 198m (650ft) long, 46m (150ft) above normal water surface, and 91m (299ft) above the power plant foundation. There is a combined 4 hectares (10 acres) of floor space in the power station buildings.

Installation of the turbines was not completed until 1961, with major uprating of the plant occurring in 1983. There are four 157,000hp, eleven 115,000hp, one 70,000hp, and one 55,000hp Francis-type vertical hydraulic turbines. There are 13 82,500kW, one 130,000kW, one 15,000kW, one 50,000kW, and one 40,000kW generators. Some machines are rated at 50Hz; however, all are operated at 60Hz.

There are also two 2400kW station-service units driven by Pelton water wheels. These provide electrical energy for lights and for operating cranes, pumps, motors, compressors, and other electrical equipment within the dam and powerplant.

The water reaches the turbines through four pressure penstocks, two on each side of the river. Shutoff gates control water delivery to the units. The maximum head is 180m (590ft).

The average annual net generation is about 3.5 billion kilowatt-hours (kWh). The maximum annual net generation was 6,463,483,000kWh in 1953, while the minimum

The mighty Hoover Dam. It's hard to picture the scale of the structure – from foundation rock to the road on the crest of the dam, the wall is 221m (726ft) high – the height of a 50-storey office block.

annual net generation since 1940 was 2,614,512,600kWh in 1965. About six million barrels of oil would be required by an oil-fired generating plant to supply the average amount of energy generated by the plant annually.

The sale of electricity pays all operational and maintenance costs of the dam. Intriguingly, it was also used to pay back the dam construction cost (plus interest at 3 per cent), something achieved by 1987.

Despite the dam being built at the heights of the Great Depression, care was taken with the dam's aesthetics. Initially, the plans for the facade of the dam, the power plant, the outlet tunnels and ornaments clashed with the modern look of an arch dam, and there was widespread criticism. Two years after construction began, Los Angeles-based architect Gordon B Kaufmann was brought in to redesign the exteriors.

Kaufmann applied an elegant Art Deco style to the entire project. One of his most significant changes was to alter the design of the towers rising from the crest of the dam wall. The emphasis, according to Kaufmann, was on 'an orderly series of small vertical shadows punctuated by the larger shadows of the elevator and utility towers.' Kaufmann also redesigned the spillways to be more streamlined, added lights to the top of the intake towers to provide a night-time effect, and redesigned the exterior of the power house. Denver artist Allen Tupper True was then commissioned to work on the design and decoration of the walls and floors, incorporating motifs of the Navajo and Pueblo tribes of the region. Many of these decorations are incorporated in the terrazzo floors used in the walkways and interior halls of the dam and power station.

The result is an extraordinary combination of massiveness lightened by

Inside one of the power houses. Some of the turbines are rated at 157,000hp. Sale of generated electricity paid off the complete construction costs of the dam and power stations by 1987.

deft decoration that makes the dam and power stations as much artwork as engineering.

Directly downstream from dam is the Mike O'Callaghan–Pat Tillman Memorial Bridge, completed in 2010. The world's highest concrete arch bridge, it is 579m (1900ft) long, with a 320m (1060ft) span. The bridge allows traffic to bypass the original highway across the dam wall, with this route now closed to through traffic.

An excellent – albeit expensive – tour is available of the dam. It includes walking through passages deep within the dam wall, and access to a grille placed part way up the dam wall, through which you can look downstream to the highway bridge.

Deep inside the dam wall. Note the terrazzo floor and the tiled walls and roof. (Courtesy Georgina Edgar)

The view downstream, taken from a grille set in the dam's face. In the foreground are the power stations, while the structures higher on the canyon walls are the spillway outlets, fed by underground tunnels. High in the sky is the highway bridge, opened in 2010.

In the late 1940s, a revolutionary new engine was developed; one that just stunned contemporary engineers. Developed for aircraft, the diesel two-stroke boasted 12 cylinders, a displacement of 41.1 litres – and developed up to 3000kW (4100hp). That engine was the Napier Nomad.

The first iteration, Nomad 1, was almost two engines in one. An axial compressor, intercooler, centrifugal compressor and two turbines formed one of the engines, while the piston diesel formed the other. The centrifugal compressor force-fed the diesel, while the diesel exhaust passed through one of the turbines. The combination drove two contra-rotating propellers using two separate mechanical drives – one from each 'engine.'

In Nomad II, the design was somewhat simplified – although I am not sure that 'simplified' is the right term for what remains a complex engine! In Nomad II, the 12-cylinder diesel exhaust drove a 3-stage axial turbine. The turbine shaft turned a 12-stage(!) axial flow compressor that supercharged the engine. The turbine shaft, however, also mechanically added power to the engine crank. Finally, a little more power was achieved by way of thrust coming from the turbine exhaust.

So why go to all this trouble? In two words: fuel economy! Or, more accurately, brake specific fuel consumption (BSFC). BSFC is a measurement of fuel consumption, expressed in terms of pounds of fuel per horsepower per hour. And the Nomad II? It could get down as low as 0.326lb per effective horsepower per hour – equivalent to a thermal efficiency of 42 per cent.

So what happened – why wasn't this engine widely adopted? Compared with jet-based turbo-props, the Nomad was heavier and larger. Many airline companies immediately dismissed it – American Airlines was typical in its concern about the complexity of the engine. "The Nomad appears to combine the complexities of both aviation piston engines and gas turbines," it deprecated.

A Nomad 1 on display at the National Museum of Flight in Scotland, UK. The engine, incredibly complex even in a time of many complex aircraft piston engines, remained just a prototype.

THE ROVER JET CARS

A serious attempt by a car maker to develop gas turbine cars for the road

The 1950s were a very different time to today. The Allies had won the war only a few years earlier; Germany and Japan were rebuilding after the destruction of the war – and furthermore, were working under quite restrictive terms set by their conquerors. Britain had optimism and confidence ... mixed with the need to overcome the terrible financial, economic and social toll that had been extracted during the war years. The confidence was largely in technical matters: the inventors of radar and the gas turbine engine – not to mention, for those in the know, the electronic computer.

It was a time for innovation and development – so meet the gas turbine Rovers. Already involved in developing the jet engine through their wartime work, it was shortly after the war ended that Maurice Wilks of Rover suggested to his brother Spencer that they should consider developing gas turbine cars. By the end of 1945, work had begun.

The first successful engine – the T5 – produced 75kW (100hp) at 55,000 rpm compressor speed. It also weighed less than a then-standard Rover car engine. The more powerful T8 engine was tested in a boat and then installed in a modified P4 Cyclops 75 car, appropriately equipped with the (real!) number plates: Jet 1.

Jet 1

Jet 1 was the world's first gas turbine powered car. Compared with a normal P4 75, the turbine car had its top removed and the bodywork smoothed – even to the extent that the rear doors were welded into place. Large air inlets were positioned on the flanks ahead of the rear wheels. It first ran on March 4, 1950. In this initial form, Jet 1 was capable of 137km/h (85mph), had a 0-60mph (0-97km/h) time of 14 seconds, and had a best fuel consumption of 6mpg (47 litres/100km). In 1952, Jet 1 was restyled slightly and was fitted with an uprated engine boasting 172kW (230hp). It then achieved a world record speed for gas turbine cars of 245.73km/h (152.69mph).

Jet 1 had a mid/rear mounted engine, was not fitted with a heat exchanger (to capture otherwise wasted heat) and used a twin-shaft design. The main engine section used a centrifugal compressor and an axial turbine. Power was taken from another turbine that drove the output shaft through helical reduction gears. The compressor operated at up to 40,000rpm (with light-up at 3000rpm) while the power take-off turbine had a maximum speed of 26,000rpm. The fuel tank was located at the front of the car; behind the front air intake was an oil cooler, the same unit used on the Land-Rover. The exhaust outlet of the turbine was directed upwards through two grilles placed on the horizontal deck behind the seats.

T3

The T3 was the first Rover specifically designed around a gas turbine engine. In October 1956, Rover produced a brochure detailing the car. From the brochure:

"Since experimental work on 'JET 1', the world's first gas turbine car, began in 1946 there has been intense activity at Solihull, headquarters of The Rover Company. In 1955 a second prototype was produced which had a rear mounted gas turbine in a normal saloon body. [This second turbine

Looks can deceive: this early 1950s Rover Jet 1 packed a mid-mounted gas turbine engine and achieved 245.73km/h (152.69mph). Photographed at the Science Museum in London.

As befits a development car, Jet 1 was well instrumented. The speedo runs to 140mph – presumably, the needle was off the scale during the record run!

car was so ugly that photos of it were stamped 'secret' and it was never publicly shown!] But now comes 'T.3', Rover's first practical, specially designed gas turbine-powered motor car.

"With the engine mounted at the rear it has been possible to design a body having a low bonnet line which, together with a deep wrap-round windscreen and large rear window, gives exceptionally safe visibility. Bodywork is of glass-reinforced plastic, while four-wheel drive and a De-Dion rear axle are included in the technical specification. Four-wheel drive is considered a desirable safety factor on a car that has such a high torque to weight ratio.

"The engine is a development of the well-known 15j60 industrial gas turbine, and consists of a single stage centrifugal compressor with a maximum speed of 52,000 r.p.m. driven by a single stage axial turbine re-designed so that it takes only sufficient power from the gas stream to drive the compressor and fuel and oil pumps. A second single stage power turbine has been added which takes the remaining power from

The T3 used a fibreglass body, all-wheel drive and a rear-mounted turbine. Fuel consumption was much better than the previous Rover turbine cars but would still not have been acceptable to the buying public. This photo was taken at the British Motor Museum, Gaydon, UK, where you can see most of the other Rover jet cars.

At maxima of 52,000rpm and 830°C (1530°F) the turbine engine developed 82kW (110hp). The exhaust ducted out of the top of the rear lid at 200°C (390°F).

the gas stream and drives the front and rear differential units. This reduction gear also incorporates a reverse gear which can be selected by a central control lever.

"A plate-type secondary surface contra flow heat exchanger is mounted on top of the engine and takes heat from the exhaust gases to heat the compressed air before it enters the combustion chamber. The exhaust is ducted at about 200°C. to a square opening in the top of the boot lid, which also incorporates an ejector orifice to ventilate the engine compartment. At 52,000 Compressor r.p.m. the engine develops 110 B.H.P. with a pressure ratio of 3.85/1, a maximum temperature of 830°C. and an air mass flow of 2 lb/sec. The self-sustaining speed of the engine is 15,000 r.p.m."

The T4 jet Rover used a front-mounted turbine engine that drove the font wheels. The body was based on the P6 (still two years away from release). Rover said at the time that the car could be ready for production in only three years – if buyers were prepared to fork out about double what any existing Rover then cost.

Fuel consumption figures were quoted as being:

Speed	mpg	litres/100km
Steady 40mph	13.8	20.5
Steady 60mph	14.3	19.8
Steady 80mph	12.8	22.1

The standing start 0-60mph (0-97km/h) time was 10.5 seconds and the 0-80mph (0-130km/h) time was 17.7 seconds – both very rapid in 1956.

Unlike Jet 1, in the T3 the engine was mounted over and behind the rear wheels, rather than ahead of them. Reduction gearing from the power turbine to the wheels took place in three stages: two helical gears sets and then the final drive that comprised a crown-wheel and bevel. The four-wheel drive system used a freewheel between the front and rear differentials.

Surely the most exciting gauge you could have on any car dashboard? Rover T4 jet car.

T4

The T4 was as close to a road-going turbine car as Rover produced. Based on the P6 2000 model (still two years away from release) but equipped with a new nose, the 2S/140 engine was located at the front and drove the front wheels. Oddly, rear suspension went from the T3 (and production P6) de Dion rear end to independent swing arms, although still with coil springs.

At the time of release, Rover claimed that the T4 could be produced in three years – if the market was ready to accept it at a price of £3000-£4000. (At the time, the most expensive production Rover cost only £1948.) However, it is suggested that even Rover realised by this time that the writing was on the wall for turbine-powered cars.

Author Graham Robson got to drive the T4. In a road test article for *Thoroughbred & Classic Cars* magazine in 1975, he wrote:

"Starting drill is simple but drawn out – turning the key actuates the special Lucas starter motor which winds away for several seconds. A faint, distant whine rises in pitch and intensity before light-up occurs and the engine settles down to 'idle' at 35,000rpm. This is enough to cause the car to creep along the road if the brakes are not applied, as there is about 4bhp residual at idle.

"To get moving, engage forward gear and depress 'loud pedal' – after a jet lag of about 3 seconds, the engine speed rises rapidly to 50,000rpm and the car whooshes off up the road leaving engine noise behind (although this is quite acceptable to passers-by). 60mph is reached in 8 secs (á la 3500S) with very civilised handling."

Rover-BRM

The last of Rover's experimental gas turbine cars was the Rover-BRM race car. A widened BRM Grand Prix chassis was used as the basis of the car, and the 112kW (150hp) engine was a development of that used in the T4. Drive from the rear-mounted engine was to the rear wheels through a gearbox that gave both forward and reverse (however, reverse gear was fitted only because race regs required it – it wasn't to be actually used). Fuelled with kerosene (the other Rover gas turbine cars could run on a variety of fuels, including kerosene and petrol), the car used a 48-gallon (218-litre) tank.

The car was entered in the 1963 Le Mans 24 hour race, where it would have finished eighth, had it been a formal entry within the existing classes. In 1964 it was fitted with a new body and the engine was modified to incorporate a heat exchanger that used ceramic discs made by Corning Glass Works of the US. However, the car did not compete in that year's race because of a lack of test time and the car sustaining damage in transit. In 1965 it ran in Le Mans as an official entry, with Graham Hill and Jackie Stewart driving. In this race, it achieved 10th place, averaging 159km/h (98.8mph).

The turbine Rover-BRM was primarily a race car – it ran at Le Mans. However, at least one road test exists of the car being tested on public highways!

THE CHRYSLER TURBINE CAR – Driving on the street in 1963!

In 1963 the Chrysler Corporation built over 50 gas turbine-powered cars. Incredibly, it then lent them – free of charge – to normal citizens to drive for three months each, with comments and feedback encouraged at the end of the period.

So, 55 years ago, a number of lucky people in the US were driving jet cars around on public roads – taking them to the supermarket, commuting to work, and probably parking them at drive-ins! They were certainly the only cars on the road that had tachos going to 60,000rpm.

Released to the press in New York on 14 May, 1963, the Chrysler Corporation Turbine Car used a body produced by Ghia with mechanicals the work of Chrysler. Except for the revolutionary driveline, the rest of the car followed typical US contemporary car practice. But that jet-like exhaust note – said by some to have been deliberately left loud – was like no other family car on the block ...

The Chrysler Corporation Turbine Car used an elegant body built by Ghia. This example is pictured in the Henry Ford museum, Michigan, USA.

The turbine engine makes its own wind. In the turbo-prop aircraft engine, the compressor (green) draws in cool air which combines with fuel to produce a hot, rapidly expanding gas–the "wind." This gas turns the turbine wheel (red)–the "windmill"–turning the propeller and compressor. The hot exhaust gases, escaping from rear, provide jet thrust.

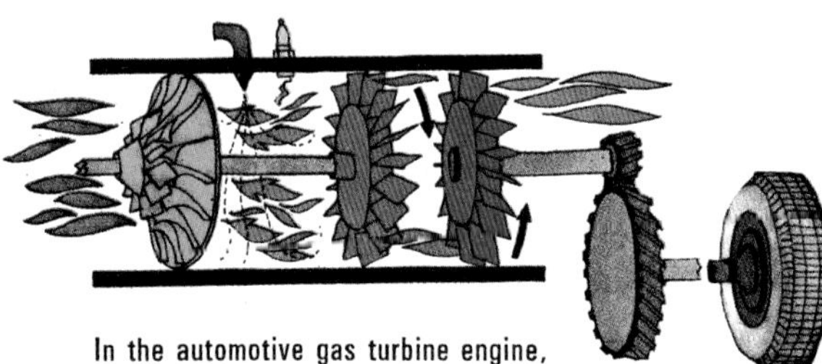

In the automotive gas turbine engine, there are two stages: 1st turbine drives the compressor; 2nd turbine drives the wheels.

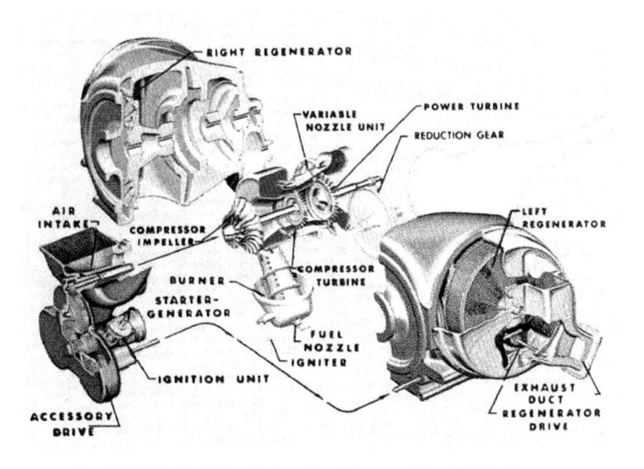

EXPLODED VIEW of Chrysler Corporation Twin-Regenerator Gas Turbine Engine showing main components.

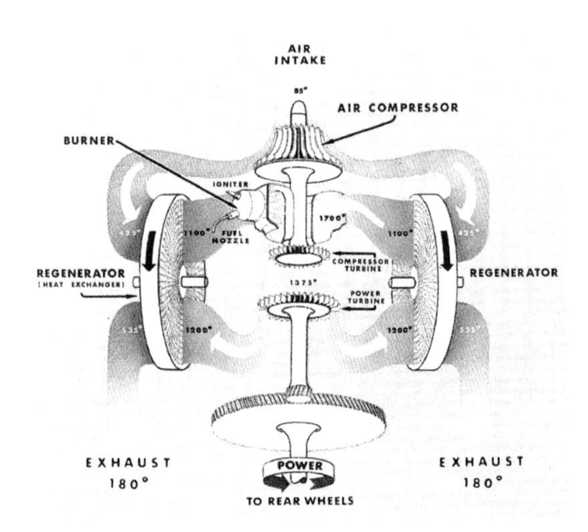

SCHEMATIC DRAWING shows direction of air flow and various temperatures from incoming fresh air to exhaust of Chrysler Corporation's Gas Turbine Engine.

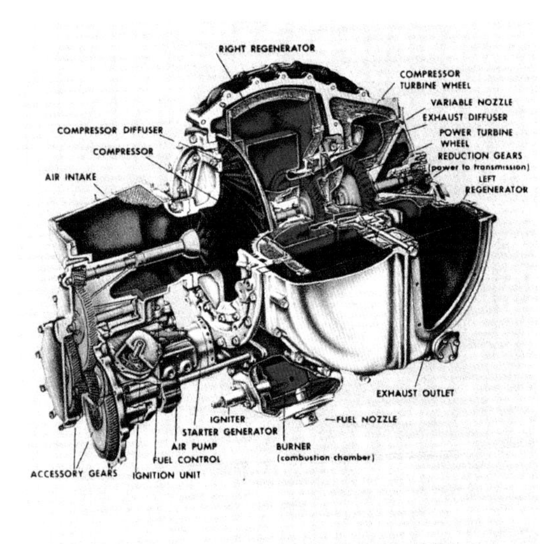

CUTAWAY VIEW of Chrysler Corporation's Twin-Regenerator Gas Turbine Engine which powers the company's limited production turbine passenger car.

An extract from the contemporary Chrysler Corporation Turbine Car pamphlet. The technical explanations were actually very good! (Owned by the author)

THE CORVAIR TURBO FLAT SIX – World's first turbo car production engine

The Corvair Spyder used the first mass-produced automotive turbocharged engine. Débuting in mid-1962, the engine developed 112kW (150hp) from its 2.4-litre (145 cubic inch) engine. While matching the specific power of only the most radical multi-carburettor or fuel injected engines of the time, it possessed a smooth idle, great drivability and excellent fuel economy. The turbo Corvair engine continued until the end of the 1966 model year, after having its displacement raised to 2.7 litres (164 cubic inches) and power raised to 134kW (180hp) in 1965.

Think 'flat six and turbo' and most people think 'Porsche.' But this turbo engine is from a 1962 Chevrolet Corvair. It's the world's first production car turbo engine. The turbo is hiding under the heat shield at the top-right. This engine was photographed at the Pierce Arrow museum, Buffalo, New York.

Chapter 5: Hovercraft and airships

The crazy brave

We live in an era where transport development has become a series of refinements, rather than invention. Consider cars, trains, aircraft and ships – even the most recent of these has been around for well over a century. So it's hard for us to imagine anyone inventing a *completely new mode of transport* – but that was the case for the two types of transport covered in this chapter, airships and hovercraft.

Both types of machines operate above the ground, but the mechanism by which they do so is quite different. Airships – differing from blimps or balloons because they use internal stiffening structures – are filled with gas that is lighter than air. Airships literally float. Hovercraft use large fans to pressurise the air beneath them (or actually, around their lower periphery). This cushion of air supports the hovercraft, allowing it to travel above the surface. If the fans stop, the hovercraft rests on the ground or water beneath it. Both types of machine use propellers for forward propulsion.

And why do these machines fascinate me? My father was always interested in airships, and so I don't remember a time when airships were unknown to me. Hovercraft – at least the big British ones – came into being when I was a child in the 1970s. I remember making my own pseudo-hovercraft from folded paper sheets when I was about five or six years old, blowing

The Hindenburg over New York.

them along with my breath and hoping they'd rise above the floor!

About that time, a large hovercraft visited Adelaide, Australia, where we lived. In 1972 my mother took me to see the hovercraft, that we found resting quietly on the beach at St Kilda, an area of flat tidal sands and mangroves very much suited to hovercraft travel. The hovercraft was deserted; we were the only people looking at it. As a 5-year-old, I was amazed at the size of the hovercraft rearing above me, and I remember that I tentatively approached it and poked the rubber skirt with my finger. The rubber was hard and stiff, and that surprised me – so I must have realised that in operation, the skirt was flexible.

Many decades passed, and our family decided to visit Germany – the birthplace of airships. Friedrichshafen on Lake Constance on the German/Swiss border was where Count Ferdinand von Zeppelin first built airships, mostly in the first few decades of the 20th century. There is a museum at Friedrichshafen devoted to his works. The museum was fabulous, and my interest in airships grew. Then, when we visited the UK, we made our way to Cardington, where the two vast airship sheds bear testament to Britain's attempts at building airships.

And hovercraft? The world's only hovercraft museum is in Lee-on-the-Solent, UK. At the time we visited, there were two of the enormous SR.N4 hovercraft (sadly, one has since been scrapped). These were the hovercraft that plied the cross-Channel trade to and from France, flying at extraordinary speed through the busy shipping lanes. Three crew sat in the high cockpit: one crew member spent all their time staring at a radar scope and directing the commander, as the vessel raced between the plodding ships. Glorious machines powered by marinized jet engines, these hovercraft carried both cars and passengers. Some left from Dover, while at Ramsgate a dedicated hoverport was established.

The Ramsgate hoverport's terminal and maintenance buildings are long gone, the site now overgrown with weeds. But it's an extraordinary evocative place: the car park's painted arrows still visible on the bitumen, and the guiding line for arriving and departing hovercraft just discernible. Look hard and you'll even find some rusting shackles used as hovercraft tie-down points.

And can you still ride in a hovercraft? You can: the hovercraft that runs from Portsmouth to Ryde on the Isle of Wight is the world's last regularly scheduled service. Did it live up to our expectations? Yes it did!

There's a common dismissiveness found when people talk about airships and hovercraft. Like, *what were they thinking? Gosh, what flops.* But that is to completely miss the bravery, excitement and engineering that these inventions represented. Of any vehicle ever built, I'd like most to travel in the Hindenburg across the Atlantic – but no, not on its last flight ...

The world's last regularly scheduled hovercraft crosses the Solent to the Isle of Wight in the UK.

SR.N4 hovercraft crossing the English Channel. Its cruising speed was 40-60 knots (74-111km/h, 46-69mph). Crossing one of the busiest shipping channels in the world, it had to duck and weave between the slow-moving ships. (Courtesy Hovercraft Musuem)

THE SR.N4 HOVERCRAFT

One of the most extraordinary vehicles ever built

The SR.N4 was one of the world's largest commercial hovercraft, and was designed for passenger/vehicle ferry operations on stage lengths up to 185km (100 nautical miles) on coastal water routes. The craft operated across the English Channel between 1968 and 2000. The discontinuation of their service was due to multiple reasons: the machines had reached the end of their useful life, and the opening of the Channel Tunnel in 1994 decreased demand. And there was another reason too. In 1999, duty-free alcohol was abolished in Europe, leading to decrease in foot passengers ducking across to France for the day to load up with cheap alcohol.

Here's the technical make-up of those wondrous machines.

Power

Power was supplied by four 3400 shaft hp Rolls-Royce Marine Proteus free turbine, turboshaft engines, located in pairs at the rear of the craft on either side of the vehicle deck. Each had a maximum rating of 4250 shaft hp but usually operated at 3400 shaft hp when cruising. Each engine was connected to one of four identical propeller/fan units, two forward and two aft. The propulsion propellers, made by Hawker Siddeley Dynamics were of the four-bladed, variable and reversible pitch type, 5.8m (19ft) in diameter.

The lift fans, made by the British Hovercraft Corporation, were of the 12-bladed centrifugal type, 3.5m (11½ft) in diameter. Since the gear ratios between the engine, fan and propeller were fixed, the power distribution could be altered by varying the propeller pitch and hence changing the speed of the system, which accordingly altered the power absorbed by the fixed-pitch fan.

The power absorbed by the fan could be varied from almost zero shaft hp (ie boating with minimum power) to 2100 shaft hp, within the propeller and engine speed limitations. A typical division on maximum cruise power would be 2000 shaft hp to the propeller and 1150 shaft hp to the fan; the remaining 250 shaft hp could be accounted for by engine power fall-off, due to the turbine rpm drop, transmission losses and auxiliary drives.

The drive shafts from the engine consisted of flanged light-alloy tubes approximately 2.3m (7½ft) long, supported by steady bearings and connected by self-aligning couplings. Shafting to the rear propeller/fan units was comparatively short, but to the forward units was approximately 18.3m (60ft). The main gearbox of each unit comprised a spiral bevel reduction gear, with outputs at the top and bottom of the box to the vertical propeller and fan drive shafts, respectively.

The design of the vertical shafts and couplings was similar to the main transmission shafts, except that the shafts above the main gearbox were of steel instead of light alloy to transmit the much greater torque loads to the propeller. This gearbox was equipped with a power take-off for an auxiliary gearbox with drives for pressure

The hovercraft's propellers were each 5.8m (19ft) in diameter. These, and complete hovercraft, can be seen at the Hovercraft Museum in Lee-on-the-Solent, UK.

and scavenge lubricating oil pumps, and also a hydraulic pump for the pylon and fin steering control.

The upper gearbox, mounted on top of the pylon, turned the propeller drive through 90 degrees, and had a gear ratio of 1.16:1. This gearbox had its own self-contained lubricating system. Engines and auxiliaries were readily accessible for maintenance from inside the craft, while engine, propellers, pylons and all gearboxes could be removed for overhaul without disturbing the main structure. The fan rotated on a pintle which was attached to the main structure.

Controls

The craft control systems enabled the thrust lines and pitch angles of the propellers to be varied either collectively or differentially. The fins and rudders moved in step with the aft pylons. The pylons, fins and rudders moved through ±35°, ±30°and ±40°, respectively.

Demand signals for pylon and fin angles were transmitted electrically from the commander's controls. These were compared with the pylon or fin feedback signals, and the differences were then amplified to actuate the hydraulic jacks mounted at the base of the pylon or fin structure. Similar electro-hydraulic signalling and feedback signals were used to control propeller pitches. The commander's controls included a rudder bar which steered the craft by pivoting the propeller pylons differentially. For example, if the right foot was moved forward, the forward pylons moved clockwise, viewed from above, and the aft pylons and fins moved anti-clockwise, thus producing a turning movement to starboard.

The foregoing applied with positive thrust on the propellers, but if negative thrust was applied, as in the case of using the propellers for braking, the pylons and fins were automatically turned to opposing angles, thus maintaining the turn. A wheel mounted on a control column allowed the commander to move the pylons and fins in unison to provide a drift to port or starboard as required.

The control of the distribution of power between each propeller and fan was by propeller pitch lever. The pitch of all four propellers could be adjusted collectively over a limited range by a fore and aft movement of the control wheel.

Hull

Construction was primarily of high strength, aluminium clad, aluminium alloy, suitably protected against the corrosive effects of seawater. The basic structure was the buoyancy chamber, built around a grid of longitudinal and transverse frames, which formed 24 watertight sub-divisions for safety. The design ensured that even a rip from end-to-end would not cause the craft to sink or overturn. The reserve buoyancy was 175 per cent, the total available buoyancy amounting to more than 550 tons.

Top and bottom surfaces of the buoyancy chamber were formed by sandwich construction panels bolted onto the frames, the top surface being the vehicle deck. Panels covering the central 4.9m (16ft) section of the deck were reinforced to carry unladen coaches, or commercial vehicles up to 9 tons gross weight (maximum axle load

I am sitting in the commander's seat of an SR.N4 hovercraft at the Hovercraft Museum. Saunders Roe, the maker of the hovercraft, had a strong background in aviation, and the control cabin is similar to contemporary aircraft. Some suggest in fact that some of the aborted Princess flying boat controls were transferred straight over. The hovercraft's servo controls were said to have come from the Saunders Roe missile program! (Courtesy Georgina Edgar)

Unfortunately, this SR.N4 has gone to the scrapyard, but her sister is still on display at the Hovercraft Museum. Note how she dwarfs the two other hovercraft in the picture.

5900kg, or 13,000lb), while the remainder were designed solely to carry cars and light vehicles (maximum axle load 2040kg, 4500lb). An articulated loading ramp, 5.5m (18ft) wide, which could be lowered to ground level, was built into the bows, while doors extending the full width of the centre deck were provided at the aft end.

Similar grid construction was used on the elevated passenger-carrying decks and the roof, where the panels were supported by deep transverse and longitudinal frames. The buoyancy chamber was joined to the roof by longitudinal walls to form a stiff fore-and-aft structure. Lateral bending was taken mainly by the buoyancy tanks. All horizontal surfaces were of prefabricated sandwich panels with the exception of the roof, which was of skin and stringer panels. Double curvature was avoided other than in the region of the air intakes and bow.

Each fan air intake was bifurcated and had an athwartships bulkhead at both front and rear, supporting a beam carrying the transmission main gearbox and the propeller pylon. The all-moving fins and rudders behind the aft pylons pivoted on pintles just ahead of the rear bulkhead. The fans delivered air to the cushion via a peripheral fingered bag skirt. The material used for both bags and fingers was nylon, coated with neoprene and/or natural rubber, the fingers and cones being made from a heavier weight material than the trunks.

Hovercraft undertook 'flights' and the crew that flew them had to be familiar with both aviation and marine vehicle behaviours.

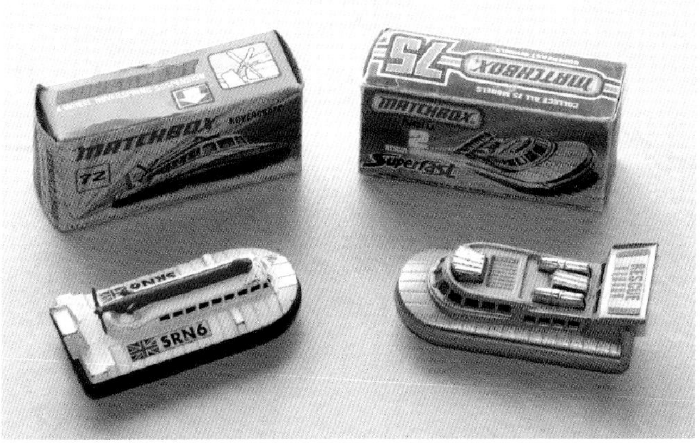

Matchbox produced hovercraft toys – just the thing for the child who wanted to keep up with technology. Unfortunately, Matchbox didn't produce a model SR.N4. (Courtesy Alexander Edgar)

Accommodation

The basic manning requirement was for a commander, and engineer radio operator, and a radar operator/navigator. A seat was provided for a fourth crew member or a crew member in training.

The remainder of the crew (ie those concerned with passenger service or car handling), were accommodated in the main cabins. The compact control cabin, which provided nearly 360-degree vision, was entered by one of two ways. The normal method, when the cars were arranged in four lanes, was by a hatch in the cabin floor, reached by a ladder from the car deck. When heavy vehicles were carried on the centre section, or if, for some other reason, the ladder had to be retracted, a door in the side of the port forward passenger cabin gave access to a ladder leading on to the main cabin roof. From the roof a door gave access into the control cabin. The craft, as configured for Channel service, carried 282 passengers and 37 cars.

Cross-channel hovercraft left from both Dover and Ramsgate; this is the 1971 cover of a pamphlet for the Dover/Boulogne and Dover/Calais service. Contemporary publicity made the most of the unique craft. (Owned by the author)

The car deck occupied the large central area of the craft, with large stern doors and a bow ramp providing a drive-on/drive-off facility. Separate side doors gave access to the passenger cabins that flanked the car deck. The outer cabins had large windows that extended the full length of the craft. The control cabin was sited centrally and forward on top of the superstructure to give maximum view.

You can explore an SR.N4 hovercraft at the Hovercraft Museum at Lee-on-the-Solent, UK.

Ramsgate hoverport

Some of the SR.N4 hovercraft operated from the purpose-constructed hoverport located at Pegwell Bay, Ramsgate. Built at a time of enormous optimism for hovercraft travel, the hoverport was seen as the first of many that would be built around the world.

Just read the original press release:

"Hoverlloyd built the world's first international hoverport on the sands at Pegwell Bay, Ramsgate, Kent, with two possible targets – to make it the smoothest, quickest possible passenger handling port, and to give passengers a comfortable and enjoyable time. The result is a streamlined Hoverport which equals the standards of any international airport – with a little more besides.

"The Hoverport cost £1,500,000 to build and can handle 1,000,000 passengers each year; it has the fastest and most direct route of any Channel port to London via the M2 Motorway.

"The long, low, handsome design of its all-in-one building fits snugly into the sheltered bay. Inside there is a shopping centre with an Exchange Bureau, a Bank

Hoverlloyd SR.N4 hovercraft loading at the Pegwell Bay, Ramsgate hoverport. The purpose-built terminal building can be seen on the right. (Courtesy Hovercraft Musuem)

HOVERLLOYD
TO FRANCE

Official First Day Cover

HOVERLLOYD
INAUGURAL FLIGHT
PEGWELL BAY
2 APRIL 1969
RAMSGATE KENT

Hoverlloyd Ltd.,
International Hoverport,
RAMSGATE,
Kent.

Craft Commander

COMMEMORATING THE FIRST SRN4
MOUNTBATTEN CLASS HOVERCRAFT
SERVICE BETWEEN RAMSGATE AND CALAIS

From the pride of the inaugural hovercraft flight from Pegwell Bay Hoverport to, nearly 50 years later ...

... all that is left of the grand dream – the desolate site of the hoverport at Pegwell Bay, Ramsgate. I found it an extraordinarily evocative place to visit.

BUT WAS CHRISTOPHER COCKERELL FIRST?

All Western descriptions of the hovercraft and its origins name British engineer Christopher Cockerell as the inventor of the craft.

Cockerell experimented with a vacuum cleaner (working as a blower) and two tin cans organised to create an annular gap. He found that the lift was far more effective if air was channelled into this gap around the base of the craft. By 1955 he had built a working model from balsa wood and then patented the idea. Cockerell then approached the UK government who placed the idea on the secret list. However, the UK government was not interested in providing funding and the idea was de-classified in 1958. Saunders-Roe (the 'SR' in the type name of many British hovercraft) then built the SR.N1, first shown in June 1959. This craft crossed the English Channel, with larger craft then developed. The first SR.N4 trials took place in early 1968, and Britain was seen to lead the world in this new technology.

But in the Soviet Union, development of air-cushion vehicles had started way back in the 1930s, most notably by Vladimir Levkov. His L-5 fast-attack boat reached 70 knots (130km/h, 81mph) in testing just before World War II. This machine did not use flexible skirts but instead used two metal hulls, one each side of the craft, projecting downwards. However, it was a true hovercraft in the way we now think of the term, being able to move over land as well as water. (*Wings of Russia* – Episode 11 of 18, available on YouTube, has good vision of this hovercraft in action.) But it does appear that the Soviet Union did not go on to create large hovercraft until after the British had done so. (Then, the USSR developed some of the largest military hovercraft in the world.)

So as with many inventions, naming only one inventor is probably misreading history.

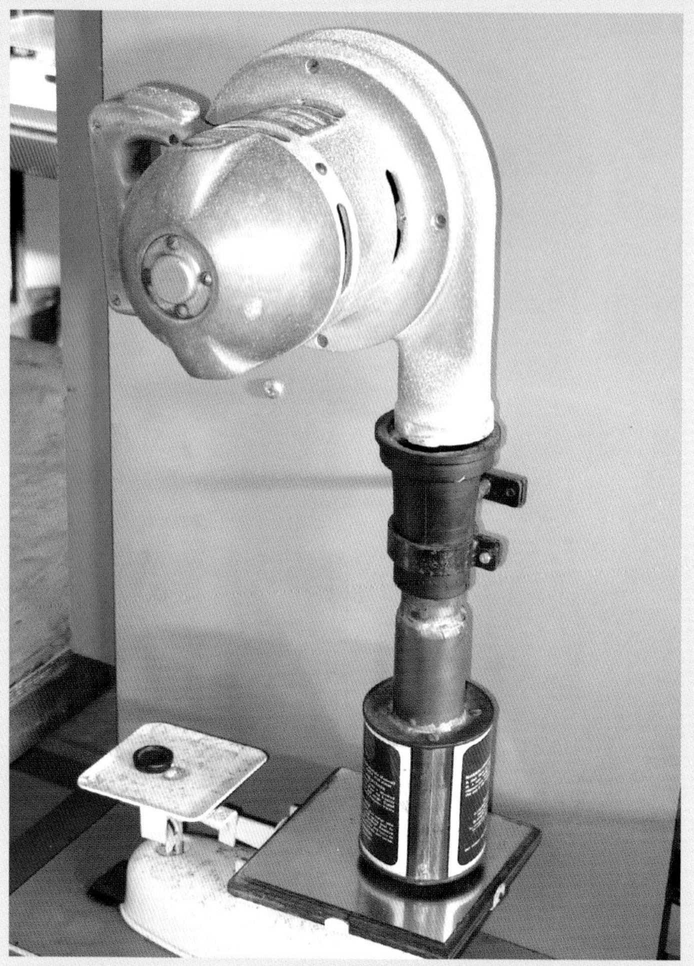

Christopher Cockerell's test equipment, complete with vacuum cleaner motor being used as a blower, the can-within-a-can and kitchen scales to measure the lift force. Photographed at the Hovercraft Museum.

and an automatic photograph machine for day passport photos.

"Around the well of the Customs Hall runs a balcony where passengers and friends can eat quite cheaply at a large Cafeteria; there is also a first-class Cocktail Bar and Restaurant where diners can watch the spectacular sight of hovercraft approaching in a storm of spray to berth on the landing platform right outside the windows. It is fascinating to see 177 tons of hovercraft sit down as gently as a feather.

"Passengers who have passed through Customs and Immigration controls have a comfortable departure lounge, with a licensed Snack Bar and a Duty Free Shop for the purchase of cigarettes, spirits and perfume."

That was written in 1970: the hoverport operated for only 18 years before all hovercraft services were centralised at Dover. The terminal and maintenance buildings then became derelict and were subsequently demolished.

The reclaimed land on which the hoverport was built is still there – and as it's open to the general public, you can walk around it. Painted lines are still visible to show the paths the hovercraft took to their loading area; the car park can still be made out, and there's a derelict pedestrian bridge that crossed one of the access roads. You can even find the steel tie-down rings for the hovercraft positioned on the maintenance pads.

AIRSHIPS

Nowadays it's the norm to consider airships as being a ridiculous dead-end technology: the Hindenburg has become a synonym for disaster. But that's a reinvention of history that ignores the overwhelming technological dominance that lighter-than-air machines had over contemporary aircraft. In the 1920s and 1930s, airships were literally the only way for passengers to speedily cross giant oceans, riding high above the sea, and in glorious comfort.

Count Ferdinand von Zeppelin started building airships in the 1890s, and his first airship, LZ1, flew over Lake Constance near Friedrichshafen in southern Germany in 1900. It had a length of 128m (421ft) and a diameter of 11.7m (38.5ft). These dimensions made it the largest aircraft in the sky – by a massive margin. (In fact, even this first airship was nearly twice as long as an Airbus A380.) As with nearly all airships, the LZ1 was filled with hydrogen. Useful lift (that is, the lift remaining after the airship has lifted its own weight) was 12,428kg (27,400lb).

The LZ1 was mildly successful: it carried five people, reached an altitude of 410m (1350ft), and flew 6km (3.7mi) in 17 minutes. However, despite making another two flights, investors were not impressed, and Graf von Zeppelin dismantled the airship and sold it for scrap.

But von Zeppelin was nothing if not determined. He started construction of LZ2 in 1905, financially supported by a state lottery and other funding sources. LZ2 was first flown in 1906. It was similar in size to LZ1, but used an improved internal structure and elevators to control pitch. (LZ1 had used a moving weight.) LZ2 made only a few flights, but its sister ship, LZ3 made many flights, and was purchased by the German Army.

However, it was World War I where the potential of the airship became startlingly obvious. Zeppelins were used for bombing raids over Britain in January 1915 and these peaked in 1916. While there was widespread British concern over the strategic advantage that Zeppelins enjoyed, in practical terms the Zeppelins were not very effective. Navigation was difficult, making them inaccurate as bombers, and techniques such as bomb aiming were in their infancy. The introduction of incendiary and explosive ammunition in 1916 also gave greater success to aircraft attacking Zeppelins, and several were shot down. The Germans responded by making lighter airships that could fly at higher altitudes – higher than heavier-than-air aircraft could then reach – but this just made the Zeppelin's bombing accuracy even worse.

Following World War I, Germany was not allowed to build airships of greater than a million cubic feet. To keep the workforce busy, the companies turned to making kitchen items from the lightweight aluminium alloys that had previously been used in the world's biggest flying craft. Reflecting the enormous advantage that the Germans had developed in this new technology, the rest of the world struggled to keep up. Despite Britain building the R34 airship – the first aircraft of any type to make a two-way crossing of the Atlantic – knowledge of airship design outside Germany was not high. (The design of the R34 was said to have been based on a German airship that crashed in Britain in 1916.) This lack of understanding is best illustrated by the end of another British airship, the R38, that broke up during final tests and fell into the Humber Estuary, all while being watched by thousands of spectators. At the time of construction, it was the biggest airship in the world.

So despite the hiatus after WWI, when in 1925 Germany was allowed to recommence airship construction, it remained a world leader. Over the next 19 years, it was to show just how this technology could be developed.

Flying first in 1928, the LZ127 Graf Zeppelin was the most successful airship ever made. It had a crew of 40 and could carry 20 passengers. While not as luxurious as the later Hindenburg, it was still vastly more comfortable than the cramped, noisy and rough passenger aircraft, that could also fly only short hops. The LZ127 had a length of 236.5m (776ft) and was 30.5m (110ft) in diameter. Useful lift was 60,000kg (132,000lb). Propelled by five Maybach engines, each with 410kW (550hp), she could fly at 128km/h (80mph). During its operating life, the

(continues on page 79)

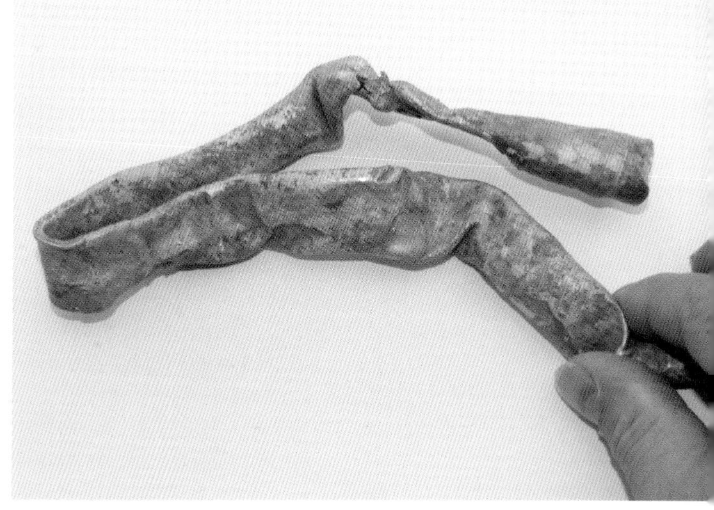

Early Zeppelin wreckage, possibly from one shot down over Britain during WWI. (Owned by the author)

The R100 in Montreal, Canada. After the disaster of the R101, the R100 was broken up and sold for scrap.

In the 1920s, Britain decided to build two airships. One, called the R100, would be built by a private company – the Airship Guarantee Company, a subsidiary of Vickers. The other, the R101, would be built by the UK government at the Royal Airship Works. The two competitive designs would each be capable of carrying 100 passengers and cargo over distances of up to 5600km (3500 miles). Both airships were considerably larger than any others of the period – including the famous German Graf Zeppelin.

The R100's design was supervised by Barnes Wallis, later to become famous as the designer of the World War II bouncing bomb and the Wellington geodesic structured bomber. (Curiously, also on the R100 engineering team was Nevil Shute Norway, later to become famous as novelist Nevil Shute.)

As befitted an airship on the cutting-edge of design, the R100 carried passengers in elaborate style. Accommodation was over three different floors, the bottom being allocated to the crew, and the upper two floors to passengers. The passenger floors were linked by a grand staircase. Cooking was carried out in an electric kitchen, and passengers were quartered in two and four-birth cabins. Power was provided by six Rolls-Royce Condor aircraft engines, running on petrol (gasoline). The R100 adopted a more streamlined approach than the previous German airships, with an overall length of 216m (709ft) and a maximum diameter of 40.5m (133ft).

In July 1930, the R100 undertook a long-distance proving flight – it flew from the UK to Canada. This flight was completed with only some minor problems – the R100 was a huge success.

And what of the government-built R101? It was plagued with problems during the build and in its first trial flights. Its performance was so bad that the decision was made to slice the airship in two to allow a new section to be inserted, increasing length by 777ft (237m) and providing greater lift. Rather than fly to Canada, the R101's long-distance proving flight was to India – but it never got there. It crashed in France, the hydrogen lift gas burning intensely and killing 48 of the 54 people on board.

Deeply shocked by the tragedy, Britain lost all enthusiasm for airships. In November 1931, the R100 was broken up.

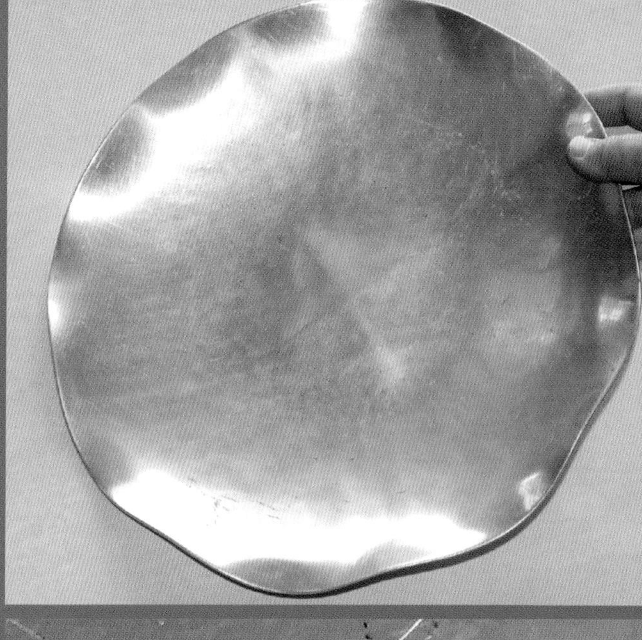

Top left: One of the controls from the British R34 airship, the first aircraft of any type to complete a two-way crossing of the Atlantic. It left from East Fortune airfield – now the National Museum of Flight in Scotland, UK, where this picture was taken.

Above: Aluminium bowl made by Zeppelin when it was prevented from producing airships after the first world war. The photo below shows the stamp on the rear. (Owned by the author)

Left: An aluminium fork, made either for use on a Zeppelin or when the company had to produce kitchenware rather than airships. The fork weighs just 28 grams! (Owned by the author)

Luftschiffbau Zeppelin LZ Friedrichshafen, Germany
No. 2

To South America by Zeppelin

1934 Time Table of the airship „Graf Zeppelin".

Friedrichshafen *	Pernambuco	Rio de Janeiro	Aeroplane connection of Syndicato Condor Ltda.		Rio de Janeiro	Pernambuco	Friedrichshafen *
Dep. Saturday evening	Arr. Tuesday evening	Arr. Thursday morning	Buenos Aires Arr. Friday	Buenos Aires D. Wednesd.	Dep. Thursday morning	Dep. Friday evening	Arr. Tuesday after noon
6. 9.	6. 12.	6. 14.	6. 15.	6. 13.	6. 14.	6. 15.	6. 19.
6. 23.	6. 26.	6. 28.	6. 30.	6. 30.	7. 1.	7. 2.	7. 6.
7. 21.	7. 24.	7. 26.	7. 27.	7. 25.	7. 26.	7. 27.	7. 31.
8. 4.	8. 7.	8. 9.	8. 10.	8. 8.	8. 9.	8. 10.	8. 14.
8. 18.	8. 21.	8. 23.	8. 24.	8. 22.	8. 23.	8. 24.	8. 28.
9. 1.	9. 4.	9. 6.	9. 7.	9. 5.	9. 6.	9. 7.	9. 11.
9. 15.	9. 18.	9. 20.	9. 21.	9. 19.	9. 20.	9. 21.	9. 25.
9. 29.	10. 2.	10. 4.	10. 5.	10. 3.	10. 4.	10. 5.	10. 9.
10. 13.	10. 16.	10. 18.	10. 19.	10. 17.	10. 18.	10. 19.	10. 23.
10. 27.	10. 30.	11. 1.	11. 2.	10. 31.	11. 1.	11. 2.	11. 6.

* In Europe there are direct aeroplane connections operated by the Deutsche Lufthansa A.-G.

The foregoing Time Table is subject to alteration, especially as regards the departure dates in and after August.

Fares:

Friedrichshafen—Pernambuco . . . . RM 1400.—
Friedrichshafen—Rio de Janeiro . . . . RM 1500.—
Pernambuco—Rio de Janeiro . . . . RM 400.—
Rio—Buenos Aires (Aeroplane) . . . . RM 400.—

Freight rates (excluding Consular fees):

Friedrichshafen—Pernambuco . . RM 8.— per kilogramme
Friedrichshafen—Rio de Janeiro RM 10.— per kilogramme

For Information and Bookings please apply to:

Hamburg-American Line,
Wm. H. Müller & Co., 66-68 Haymarket, London SW1
their agencies, travel bureaus, or:

HAPAG

Passenger and freight rates card for the Graf Zeppelin on its South American route. (Owned by the author)

Accurate re-creation of the passenger lounge of the Hindenburg, photographed at the Zeppelin Museum, Friedrichshafen, Germany. Weight needed to be kept low and the featherweight chairs in the lounge used aluminium tubing.

Passenger cabin re-created, Hindenburg. Passengers were not expected to spend many waking hours in their cabins, and so the luxury was concentrated on the lounges. Note the superb, drilled aluminium ladder – everything was light! Photographed at the Zeppelin Museum.

Original air mail envelope carried by the Hindenburg. Note the hand-written inscription at left – via Hindenburg Luft Post. (Owned by the author)

Another envelope carried by the Hindenburg – but this one on its last trip. At least 360 letters and cards (of more than 17,000 items of mail on board) reached their destinations. The charred envelope was photographed in the National Postal Museum, Washington DC, USA.

airship made 590 flights covering more than 1.7 million kilometres (over 1 million miles). It operated commercial transatlantic flights, circumnavigated the world – and all without a single passenger injury. Incredibly, it also operated a service from Germany to Brazil, taking 68 hours to do the trip.

With the success of the LZ127, work began on the Hindenburg. Designed specifically for luxurious intercontinental travel, the Hindenburg was even larger than the previous airships, with a length of 245m (804ft) and a diameter of 41m (135ft). (It was over three times the length of an Airbus A380!) It was powered by four Daimler-Benz DB 602 diesel engines, each with 890kW (1200hp). In 1936, the Hindenburg made ten return trips to the US and seven to Brazil. More than 70 passengers could be carried, with luxurious lounges, crates of wine, chef-prepared meals, and even a sealed smoking room.

However, on May 6, 1937, it all came to an end. While docking at Lakehurst in the US, the airship caught fire and was destroyed. Dramatic and shocking footage was captured by movie cameras, and was soon playing around the world. Thirteen passengers, 22 crew and one person on the ground died.

A sister ship to the Hindenburg, the LZ130 (also called Graf Zeppelin) was being built at the time of the disaster. It first flew in 1938 but never carried passengers, and was broken up in 1940.

An obvious contributing factor to the devastation that accompanied airship accidents was the flammability of the hydrogen with which they were filled. The alternative gas – helium – was not flammable, although it had a slightly lower lifting capability. However, the US held the main reserves of helium, and would not sell it to Germany. Another aspect that made airship travel dubious was the fragility of the craft. They had to steer around storms – rather than climb over them – and without weather radar or satellite imagery, this could be problematic. If you read detailed accounts of the most successful long-distance airship flights, invariably repairs had to be carried out, and near-misses during docking and departure were common.

This section of framework from the LZ 130, sister ship to the Hindenburg, shows the incredible lengths to which the designers went to achieve a super lightweight, stiff and strong structure. Photographed in the Deutsches Museum in Munich, Germany.

But, to be fair, no doubt a lot of these latter shortcomings applied also to the fixed wing aircraft of the time.

But what was indisputable was the extraordinary comfort of the airship. A pencil could be stood on end on a table for hours, and the smooth departure meant that passengers were often unaware of when the airship had left the docking tower. The open windows of the observation lounge, and the quietness of the airship, meant that dogs could be heard barking on the ground. Airships were the largest structures for their static mass ever made, and especially in the case of the big German airships, were technically sophisticated and crewed at a very high level of competence. If helium had been widely available in the 1930s, I think the story of the airship could have had a quite different ending.

Elegant cup and saucer from a later Zeppelin. LZ stood for Luftschiftbau Zeppelin – literally translated as 'airlift Zeppelin.' Photographed in the Science Museum, London.

One of the five engine cars from the LZ127 Graf Zeppelin. The engine is a Maybach V12 developing 419kW (570hp) at just 1600rpm. Capacity is no less than 33.3 litres (2032 cubic inches)! Photographed at the Zeppelin Museum, Friedrichshafen, Germany.

Acknowledgement: Some of the text of this chapter is based on material drawn from the Hovercraft Museum and is used with permission.

Chapter 6:

Cars

Engineering and society

I've loved cars all my life. I think they are technically fascinating, artistic expressions of the society in which they were born. And perhaps no place in the world better reflects that than the Mercedes-Benz museum in Stuttgart, Germany.

Mercedes understands its own history better than anyone, and the museum is supreme in displaying technical innovative and an uncompromising attitude towards excellence, mixed with a frankness about the company's past that is surprising and welcome. The museum is divided into floors connected by ramps – you start on the top floor, and work your way downwards. Each floor covers an important era of automotive design, with the models chosen to reflect that period. The connecting ramps carefully establish – by means of photos and small exhibits – important social and historical influences of the periods. And the company doesn't step away from its use of slave labour in its factories in World War II – this dark past is clearly and objectively described.

Visiting car museums around the world makes you realise the bias that you unconsciously develop when it comes to reading history. For example, until I went to the

Changing perceptions – this is the 1925 Miller Junior Eight, probably the most advanced racing car in the world in its day. It used a 2-litre straight eight which drove the front wheels, and could achieve 225km/h (140mph). Photographed at the Petersen Automotive Museum, Los Angeles, USA.

Louwman Museum in The Hague, Netherlands, I would have bet that the world's first car with a six-cylinder engine, all-wheel drive and four-wheel brakes would have been German ... but I'd have been wrong. The Dutch Spyker 60hp racing car achieved all of that – and did so way back in 1903, only 17 years after the very first 'real' car was produced by Carl Benz.

Looking at these very old machines, I am always struck by their sophistication. One tends to think of these cars as being rather primitive – sort of blacksmith, backyard jobs. But in fact when viewed close-up, these cars show very careful design and highly professional construction. For example, even very early cars show excellent steering linkage design (giving Ackermann steering angles) and even the first car used rack and pinion steering. Steam cars used automatic and sophisticated control systems, reducing driver work.

Those early machines are utilitarian in styling, however it wasn't long before styling started to make an appearance, with shapely, flowing fenders, and headlights that appeared streamlined (even if they weren't!). The Cord L-29 of 1929 reflected this, but it was the next model Cord – the 810 – that really established automotive styling as an artform in its own right. Incredibly, just six years separated the two models.

And these flowing, curvaceous lines were also able to be seen in some racing machines of the 1950s. Before appendages to create aerodynamic downforce were necessary, racing cars were often styled for low aerodynamic drag. The W 196 Mercedes of the mid 1950s was available in two forms – a pretty open-wheel version, and a fabulously beautiful fully-streamlined version. In an era where spaceframes were used rather than monocoques, and where, with the chassis unclothed, all the mechanical parts can be seen, these are fascinating cars to look at closely.

Then from beautiful back to utilitarian – the 1960s Mini. The first Mini was released in 1959, but its impact – in both popular culture and car design – were felt through the 1960s. Curiously, the car was very nearly not a success, with initial sales low – a good reason not to equate sales numbers with engineering significance. The Mini shown in this chapter is a full cutaway, and these pictures allow you to marvel at the extraordinary space efficiency achieved by designer Alec Issigonis. What is normally not so remarked upon is the compact and clever suspension.

And then we come to the near-present, with the ground-breaking hybrid Prius and the shape of the future in the Tesla Model S.

1890s – BENZ PATENT MOTOR CAR

On July 3, 1886 the Benz Patent Motor Car – regarded as the world's first car – completed its maiden journey in public.

Engine

The heart of the Benz Patent Motor Car was a single-cylinder, four-stroke engine with a displacement of 954cc (58 cubic inches). This design already incorporated several of the key features found in most internal combustion engines of today, including a crankshaft with counterweights, electric ignition and water cooling. This unit developed a peak output of 0.55kW (0.7hp) at 400rpm.

The engine weighed only about 100kg (220lb) and used an open crankcase. The intake slide valve was controlled by an eccentric rod, the exhaust poppet valve being operated via a cam disc, rocker arm and pushrod. Drip-feed lubrication was used. Benz's design had the large flywheel mounted horizontally on the chassis, because he feared the gyroscopic effect of a vertical arrangement would interfere with the steering and stability of the vehicle.

Preparation of the fuel-air mixture was handled by the 'surface' or evaporative carburettor, another Benz design, which also doubled as a 4.5-litre (about 1 gallon) fuel reservoir. The vehicle fuel economy was 10 litres/100km (25mpg). The fact that fuel capacity was inadequate for longer distances was not initially a major concern. First and foremost, the purpose of the Benz Patent Motor Car was to demonstrate that the overall design – and hence the concept of the automobile – was fit for purpose.

The precise composition of the petrol-air mixture was controlled by a sleeve valve, the position of which determined power output by opening or closing holes for the auxiliary intake duct. This sleeve valve was easily accessible beneath the driver's seat. The engine was started with a swing of the flywheel. Benz spent a great deal of time experimenting with the ignition before he eventually came up with a solution that suited the low battery voltage of the day. He transformed it up to a higher voltage using a spark coil. The spark plug was also one of his own designs.

Cooling the internal combustion engine presented a particular problem, since unlike a stationary engine, it could not merely be attached to a cold water supply housed inside a building. Benz opted for simple evaporation cooling, which proved highly effective and more than adequate for the low engine output. The steam

The world's first motor car. All photos of this car show an accurate replica. (Courtesy Mercedes)

The vehicle was an innovative mix of bicycle manufacture (the wheels), cutting-edge engineering (the engine) and horse-drawn vehicle design (the springs). (Courtesy Mercedes)

Because these vehicles are replicas, it's hard to know how whether aspects like machining quality match the original. On the replicas at least, quality is high. This, and the other detail photos shown here, were taken at the Petersen Automotive Museum, Los Angeles, USA.

Large wheels at the rear and fully elliptic leaf springs must have given an acceptable ride, even without pneumatic tyres and on poor roads.

produced was allowed to escape, meaning that water consumption during a journey was considerable. More efficient closed-loop cooling systems were not developed until around ten years later, when increased engine outputs and the higher demands placed on a vehicle's reliability made this a necessary improvement.

Chassis and body

The vehicle frame of the Patent Motor Car was shaped and welded together from tubular steel. Benz initially opted for a single steered front wheel, thus creating a three-wheeled vehicle. The front wheel was mounted in an unsprung fork, and was turned using a lever attached to a toothed rack. Not until several years later, in 1893, did Benz invent axle-pivot steering for two steered front wheels, one of the most significant milestones in automotive design.

Benz produced the three wire-spoke wheels with solid rubber tyres, buying in only the rims. As was typical of bicycle designs of the day, the front wheel was mounted on a ball bearing, the rear wheels in white alloy bushings. The car was driven at the rear wheels by two chains to the right and left of the through-drive countershaft; the rear wheels were attached to the frame via a rigid axle and fully elliptic springs. It was not by chance that the Patent Motor Car bore certain similarities to the bicycle: not only was the bicycle considered state-of-the-art, it also gave Benz inspiration for a vehicle that was both stable and lightweight.

The world's first car had just one speed on the countershaft in the form of a fixed drive disc with integrated differential and adjacent idler disc. There was no multiple speed transmission and no reverse gear. Thanks to the idler disc, the flat belt linking the engine and the countershaft simultaneously served as a clutch. To slip the vehicle into gear in preparation for moving off, the belt was simply shifted from the idler disc to the fixed drive disc. The reservoir for cooling water was located above the engine. The vehicle was braked using a hand lever that acted on the countershaft belt disc. The leather-upholstered seat bench was attached directly to the frame and mounted on elegantly contoured springs. A low, leather-bound rail at the back and sides provided additional support.

Test drives

For reasons of secrecy, preliminary short test drives were held at the factory premises in 1885. The first outing on the open road – staged at night – lasted only a few minutes, since the car came to a standstill after a hundred metres. But before long, one hundred metres turned into one thousand, and with each test the distance got bigger.

In his memoirs, Benz recalled: "I probably reached a speed of 16km/h (10mph) with the car. My confidence

Rack and pinion steering – yes, on the world's first real car!

grew with each outing, but on each occasion I also grew to recognise new characteristics of the engine; on the other hand, every journey showed me new improvement possibilities, so that by January 1886 I was ready to apply for a patent for the car." For the first public outing on 3 July 1886, an event documented in a press article, his son Eugen ran alongside the vehicle holding a bottle of fuel, "in order to fill up the tank should the gasoline run out."

Carl Benz applied for a patent for his Motor Car on 29 January 1886. The German Patent No 37435 is considered the birth certificate of the automobile. It bears the heading: "Vehicle with gas engine operation," and the text begins with the words: "The present design serves for the operation of mainly light vehicles and small boats, such as those used for the conveyance of one to four people. [...] For whichever system, power is provided by a small gasoline engine. This is fed with fuel from an apparatus carried on board, in which gasoline derived from ligroin or other volatile substances is produced. The engine's cylinder is maintained at a constant temperature by the evaporation of water." A French patent was applied for on 25 March the same year.

The Benz Patent Motor Car No I model demonstrated that Carl Benz's idea was viable. Benz used this experience to build a number of other models in quick succession; although these did not differ fundamentally from the original, they featured a number of minor improvements.

The No II model was also designed originally as a three-wheeler, and subsequently converted experimentally to a four-wheeled vehicle. This car, which also featured a prototype axle pivot steering system, represented a further step towards the modern car. It is thought only one example was ever built.

The No III model was improved sufficiently to be sold

in small numbers – around 25 units in total. Depending on the variant, each was equipped with an engine with an output of up to 2.2kW (3hp). In addition, the No III model featured wooden spoke wheels, a small petrol reservoir and a leather-clad, hand-operated block brake that acted directly on the rear wheels. There were also two forward gears, achieved by means of an idler disc and a two-stage fixed drive disc.

1920S – WHEN STEAM LOOKED LIKE IT COULD RULE THE ROAD

Why use an 'explosion' engine when you can have steam?

Imagine a car engine with so much torque that it needs a final drive ratio of only 1.5:1. A car engine which, at a road speed of 205km/h (127mph), is spinning at only 630rpm.

BERTHA BENZ – An extraordinary drive

It was early one August morning in 1888 when a very determined woman set out on an adventure that would change the course of history: Bertha Benz, the wife of motor car inventor Carl Benz, embarked on the first ever long-distance journey in automotive history in a Benz Patent-Motorwagen, accompanied by her two sons.

The route led from Mannheim to Pforzheim, the birthplace of Bertha Benz. It was a 106km (66mi) journey fraught with uncertainties and challenges that called for great courage on the part of the pioneering trio. The venture paid off, though, as the Patent-Motorwagen proved itself to be a perfectly suitable and flexible means of transport for longer journeys, too.

The mother of four told her husband nothing of her intentions to drive to her birthplace of Pforzheim, to visit relatives with her sons, Eugen and Richard. A concrete plan started to take shape when the school holidays started, and very early one morning, the mother and her sons sneaked the latest version of the Patent-Motorwagen out of the workshop without Carl Benz suspecting a thing.

The family pushed the three-wheeled motor car for the first few metres as they didn't want to start the engine until they were out of earshot of the house, so as not to wake the inventor. When Carl Benz later entered the kitchen, he would find a message telling him about the journey they had set out on to Pforzheim. Carl Benz's wife kept him informed of the journey's progress by sending a series of telegrams en route.

The drive from Mannheim to Pforzheim and back wasn't just a watershed moment for his Patent-Motorwagen, but for the future of the motor car...

Bertha Benz was a strong-willed, energetic woman. She encouraged her often self-doubting, obstinate and sullen husband, Carl Benz, in her own unique way, pushing him to continue time and time again after setbacks, and stood by him for nearly 60 years. She was a woman who shared her husband's far-reaching technical visions completely, and made many sacrifices as a result. She was also the world's first long-distance driver. (Courtesy Mercedes)

An engine that can be permanently connected to the wheels – no clutch or automatic transmission needed. An engine that can propel a car smoothly and effortlessly in a perfectly linear rush of power from stationary to maximum speed, all just by opening the throttle. An engine that can burn almost any combustible liquid or gaseous fuel without engine-destroying detonation and with potentially very low emissions.

It sounds a dream powerplant – but it existed and was being widely used in cars in the first few decades of last century. Steam-powered cars could do all these things – and also set the 1906 outright land speed record.

Rather than using steam turbines (see Chapter 4), the steam cars of the 1920s and 1930s used reciprocating engines. These were typically double-acting (that is, steam pressure was used to push the piston both up and down)

SPYKER 60HP RACING CAR – A 6-cylinder racing car with all-wheel drive

The 1903 Spyker 60hp was the first car in the world to be equipped with a six-cylinder engine, the first petrol-driven car with four-wheel drive, and the first car with a braking system connected to all four wheels.

The racing car was commissioned by Jacobus Spyker for the Paris to Madrid race of 1903. A young Belgian engineer, Joseph Valentin Laviolette, already had a design for an engine with six separate cylinders, and was now able to build it. He also designed a transmission that drove the front as well as the rear wheels, and fitted a transmission brake. However, the car was not ready in time for the race and was launched in December 1903 in Paris.

Lots of 'firsts' that in some cases would take decades to again be implemented – the first car with a six-cylinder engine, the first petrol-driven car with four-wheel drive, and the first car with four-wheel braking. The photo was taken in the Louwman Museum, The Hague, Netherlands.

This Stanley Steamer is a 1921 model. The boiler is located under the hood (bonnet) and with its front-mounted condenser looking much like the radiator of an internal combustion car, there is little to differentiate the steam car in appearance to its contemporaries. This car was photographed at the Haynes International Motor Museum, Somerset, UK.

Steam cars were regarded as easier to operate than internal combustion engines. For example, there was no need to adjust the ignition timing on the move, as was the case with internal combustion engine cars.

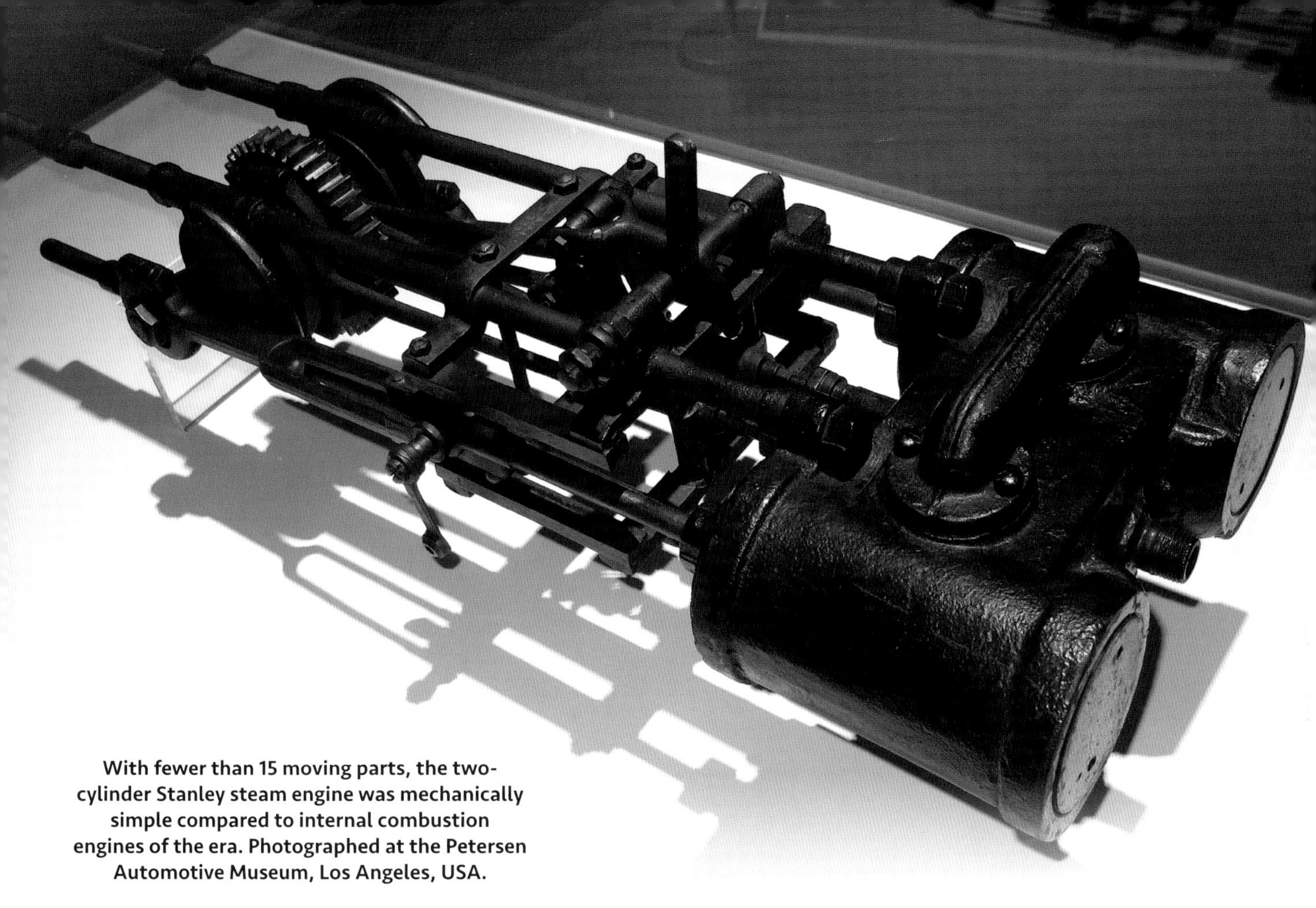

With fewer than 15 moving parts, the two-cylinder Stanley steam engine was mechanically simple compared to internal combustion engines of the era. Photographed at the Petersen Automotive Museum, Los Angeles, USA.

and had multiple cylinders. In some cases, the steam did its work twice, firstly in a high-pressure cylinder, and then secondly in a low-pressure cylinder. The valve gear included the ability to vary the 'cut-off,' allowing the stepless tuning of the engine operation for power or economy. (Cut-off describes the point in the piston stroke when steam is no longer admitted. Cut it off early and the steam continues to expand and so drive the piston, but with less consumption of steam. Make the cut-off later and more steam is used, but the power is greater.) In addition to variable cut-off, a throttle valve was used to alter the pressure of the steam entering the engine.

The use of multi-cylinder, double-acting reciprocating engines gave a high number of power pulses per turn of the crankshaft. In fact, a contemporary advertisement suggested the four-cylinder steam engine had the same intrinsic smoothness as a V12 internal combustion engine. (Incidentally, the steam car manufacturers always referred to their competition as having 'internal explosion' engines!)

While some of the boilers of these steam cars (including the pictured Stanley Steamer) were constructed on lines similar to a railway locomotive, the more advanced ones were not. In these cars, rather than having a large vessel filled with water through which tubes carrying the hot gasses from the fire passed, the approach was reversed, and water passed through the tubes. A very long coil of tube carried the water and the hot gasses from the burning fuel enveloped the tube.

Two approaches were used: in some cases, the tubes were filled with water which thermo-syphoned through the tubes, with steam collecting in a separate receptacle at the top of the system, and in other cases the water passed only once through the tubing, turning to steam as it did so. These latter designs were called 'flash steam' (as in, the water flashes into steam) and allowed very quick start-up times. In fact, from cold, flash-steam boilers could build pressure in only 30 seconds or so. However, the downside of a flash boiler design was a lack of reserve power; in a flash boiler the steam wasn't stored – it was used as quickly as it was produced, and so, if more was needed, more had to be instantly produced.

In both water tube boiler designs the risk of a major explosion was very small. This is because the volume of hot water that would instantly turn to steam when the pressure dropped (eg because of a rupturing vessel) was vastly less than in even a small fire-tube boiler. Speaking of pressures, those achieved in the water tube boilers could be much higher than in a fire-tube design. This is because the size of the pressure vessel is so much smaller – rather than a boiler measuring feet in diameter, the pressure was held by tubes perhaps only a half-inch in diameter. In fact, steam car pressures of 51bar (750psi) at 370°C (700°F) were common. (The Land Speed record-breaking Stanley Steamer of 1906 used 69bar (1000psi!)).

Burners were surprisingly sophisticated. Often rather like large Primus-style camping stoves, they used forced draught (provided by a fan run by an electric motor), venturis to draw through the fuel, and automatic control. The automatic control was on/off: as steam pressure fell, the burner was switched on and stayed on until steam pressure reached the appropriate level, whereupon the burner was extinguished. Obviously, when no steam was being used, the burner stayed off. A pilot burner was used in some cars, while others used a spark plug and ignition coil to light the mixture. The water feed was also automated; in the case of flash boilers, this had to be regulated very precisely.

In nearly all cars, the engine was mounted at the rear of the car under the floor, being mechanically integrated with the rear axle. The boiler was located at the front of the car and a condenser (which turned the steam back into water after it had done its work) took the place of the radiator. Tanks for water (because some water was inevitably lost), burner fuel, pilot light fuel and lubricating oil were distributed around the car. Pumps were also needed for the fuel, feed water and oil.

In the days when internal combustion engine cars had controls for ignition timing on the steering wheel (and it was a control that needed a lot of adjustment!), and when starting handles, gear levers, clutch pedals, lubrication reservoirs and the like all needed attention, the steam car was regarded by its makers as being simple to operate. However, even going on the contemporary instructions provided by these manufacturers, the start-up and maintenance procedures look complex. It was also apparently impossible to prevent the water freezing in sub-zero temperatures – no anti-freeze existed that could be used in the application.

The first commercial steam cars appeared around 1900 and in Germany steam trucks were still being built in the late 1930s. However, post-World War II, no new steam vehicles have been successfully commercially developed. Since then, steam cars have been the province only of amateur experimenters – despite some university-backed research projects and a few half-hearted projects by major manufacturers that took place when emission legislation was first being introduced in the US.

1930S – THE SOPHISTICATED CORDS

Eight cylinders and front-wheel drive

Looking at the Cords it's easy to develop the completely wrong idea of their origin. You know, something along the lines of: these were fantastic and ground-breaking cars from a revered manufacturer, Cord being like Mercedes-Benz of the time. And cars might, in fact, be fantastic and ground-breaking, but the business was about as far from being Mercedes-Benz as it's possible to be!

Erret Lobban Cord was in his late 20s when he was running a car dealership in Chicago in 1924. The Auburn car company offered him $36,000 a year to help put the firm back into good shape – but Cord refused, instead seeking 20 per cent of any profit he could make, plus an option to purchase stock. Within a year, he owned Auburn outright, and before long he'd also added Duesenberg cars, American Airlines, the Stinson aircraft company and the Lycoming engine plant. (And how did he do all of this? One source suggests he dumped common stock in large quantities until its value was so low that he could purchase a controlling interest for a song. A 'holiday' he took to England in 1934 was actually to avoid a government investigation into his business practices!)

It was in 1929 that a model bearing the name of Cord first appeared. This was the L-29 – a car boasting front-wheel drive, a de Dion front axle on quarter-elliptic leaf springs, and inboard hydraulic front brakes. At the back, suspension comprised a beam axle on semi-elliptic springs. Unusually, Houdaille-Hershey shock absorbers were fitted both front and back. The engine was a straight-eight, all 4935cc of it.

Cord publicity described the car:

"Docile and dignified in town, yet capable of unusual speed on the open road, the Cord commands your attention ... the principles employed in its front-wheel-drive unit have withstood the rigorous test of racing, whilst its low, sweeping lines suggest amazing acceleration and power."

However, power was only 93kW (125hp) and with the car weighing over 2 tons, performance was lethargic – even at the time. Acceleration from 0-97km/h (60mph) took over 30 seconds, and top speed was barely 121km/h (75mph).

And they weren't the only negatives. The front driveshaft CV joints wore quickly, and the weight distribution (more than half was on the back – undriven – wheels) meant the traction on loose surfaces was poor. But the styling was superb. The L-29's wheelbase was an enormous 3.5m (137.5in), and the flowing bonnet and mudguards looked magnificent.

Only 4400 units were made. When new, the car cost

The Cord L-29 used a straight eight engine and front-wheel drive. It was low, flowing and looked fast – but wasn't. The photo was taken at the Pierce-Arrow Museum in Buffalo, New York.

$3295; just five years later, the car was valued in a contemporary used car guide at $145!

The next car to be called a Cord was the 810 – one of the most beautiful cars of all time. Radically different, not only from other cars of 1936 but light-years ahead of the L-29 still on sale only four years earlier, the 810 also had front-wheel drive and an eight-cylinder engine. But this time the engine was of V8 configuration, with the engine located behind the front axle line and the gearbox ahead of it.

The shorter engine and better weight distribution (the previous L-29 straight-eight had both engine and gearbox located behind the front axle – one reason for the incredible bonnet length) meant the 810 was a sweeter car in fundamental design.

The Lycoming V8 had a capacity of 4730cc and produced 93kW (125hp) at 3500rpm. Add a Schwitzer-Cummins centrifugal supercharger, and power jumped to 127kW (170hp) at 4200rpm. A higher boost blower was then fitted that brought power up to 142kW (190hp). The unblown Cord could do 145km/h (90mph) and get to 97km/h (60mph) in 20 seconds. The supercharged version would do nearly 177km/h (110mph) and get to 60mph in 13 seconds, making it one of the fastest pre-war American production cars.

Suspension was also a step up over the L-29. Rather than the front de Dion solid axle, the 810 used independent front trailing arms with a single transverse leaf spring and an anti-roll bar. Rear suspension retained the beam axle and semi-elliptic springs.

But while the mechanicals might have been innovative,

The L-29 used a de Dion front axle mounted on quarter-elliptic leaf springs. It had inboard hydraulic front brakes. This was probably the most sophisticated front end of any car then on sale, and drew heavily for inspiration on the Miller Straight Eight racing car shown at the beginning of this chapter. Photographed at the Pierce-Arrow Museum, Buffalo, New York.

The Cord 810's styling was astonishing – from its coffin-shaped nose to its retractable headlights. Other body styling highlights included concealed hinges, a petrol filler hidden beneath a hinged flap, and a rear-hinged bonnet. This is a 1935 Cord Beverly, photographed at the Haynes International Motor Museum, Somerset, UK.

Front suspension used independent trailing arms with a single transverse leaf spring. Note the simple pull-rod that connects the spring and the arm, and the use of an anti-roll bar – the latter almost unheard-of at the time.

One of the best views of the Cord. The company did not have a sufficiently large body press to produce the complete roof panel, so it was welded together from multiple panels.

Supercharged Cord 812, built in 1937 and so one of the last produced. It was photographed at the Sinsheim Technic Museum in Germany.

this was a car that sold on style and features. The new Cord had retractable headlights, adapted from Stinson aircraft landing lights. It had a streamlined nose with wraparound radiator louvres replacing a traditional grille. Other standout features of the styling were twin streamlined tail-lights, no running boards, concealed door hinges, a petrol filler hidden beneath a hinged flap, and a rear-hinged bonnet (hood). But that wasn't all! You could also add dimmable instrument lighting, variable speed windscreen wipers, sound insulation, and a radio as standard.

But, as with the company itself, all was not quite as it seemed. To save time and money, only two dies were used for the doors of the four-door model – so the front door on one side was the same as the rear door on the other side! There was not a sufficiently large body press to produce the roof panel, so this was welded together from multiple panels. The interior handles were acquired at a bankruptcy sale, and were then equipped with large decorative plastic knobs, and the dashboard instruments were also a job lot.

But the biggest problem was the transmission. Designed to an advanced specification, the transmission used a pre-selector approach, where the driver, using a tiny lever, selected the next required gear. Then, when the clutch was pressed, that gear automatically engaged. And the problem? At the time of the car's first public display at the 1935 New York Show, the transmission wasn't finished. (In fact, the transmission was never fully sorted.) Perhaps unfortunately, the 810 was an immediate and enormous success at the New York show, with many orders taken. Bolstering consumer confidence was the apparent availability of 100 cars – but these cars were complete only externally!

The 125-inch wheelbase 810 came in four models – the Westchester and Beverly sedans (the main differences being internal trim), the Sportsman two passenger two-door, and a four passenger convertible (called a phaeton sedan). In 1937, the model designation changed to 812, and two 132-inch wheelbase models were added – the Custom Beverly and the Custom Berline.

Only 1174 Cord 810s were produced – about one-twelfth of what was expected – and 1146 Cord 812s were built. E L Cord's empire collapsed in 1937 – but it did leave behind what are among the most beautiful cars ever produced.

1950S – MERCEDES W 196 GRAND PRIX CAR

A straight eight and desmodromic valve operation in this F1 winner

The Mercedes-Benz W 196 R designed for the 1954 season met all the demands of the new Grand Prix formula set by the sport's governing body. These rules required an engine capacity of 2500cc (750cc when supercharged), a free choice of fuel mixture, and a racing distance of 300km (186mi) or a minimum of three hours.

Two versions of the car were built: streamlined, and more conventional in appearance with exposed wheels. The streamlined version was completed first because the Reims race, that started the season, permitted very high speeds.

Fritz Nallinger was in charge of the project as a whole, assisted by Rudolf Uhlenhaut, Chief Engineer of the racing department since 1936, and after the war also head of the Car Testing department.

The tubular space frame was light (36kg, or 79lb) and stiff. Giant, turbo-cooled (and at first centrally arranged) Duplex drum brakes were fitted. The 2.5-litre, naturally-aspirated eight-cylinder in-line engine was installed in the

The Mercedes W 196 Grand Prix car was built in two versions – a streamlined car (left) and a more conventional open-wheel version. (Courtesy Mercedes)

Stripped of its bodywork, the straight-eight engine looks much larger than its 2.5 litre capacity. It revved to 8500rpm and could deliver 216kW (290hp). Note the massive inboard drum brakes at the front. (Courtesy Mercedes)

space frame at an angle of 53 degrees to the right to lower the centre of gravity and reduce frontal area. In 1954 it had 191kW (256hp) at 8260rpm; by 1955 this had risen to 216kW (290hp) at 8500rpm. The engine used a built-up crankshaft with one-piece con-rods. Stroke was 68.8mm (2.7in) and the bore was 76mm (3in).

The two versions of the W 196 R were interchangeable quite effortlessly. Chassis number ten, for instance, was entered with open wheels in the 1955 Argentinian Grand Prix and the Dutch Grand Prix, and then in fully streamlined form performed tests in Monza. The body form used depended on the circuit, the strategy chosen and the likes and dislikes of the respective driver.

The W 196 R featured rear suspension swing axles with a low pivot point instead of the customary de Dion layout – a configuration justified by Uhlenhaut for its better behaviour under acceleration. Rear roll centre height was 150mm (6in). Longitudinal location of the rear hubs was by Watts linkages and springing used torsion bars. Front suspension also used torsion bars, but this time with double wishbones. Telescopic dampers were fitted front and back.

Unlike current practice, the car was designed to have a high polar moment of inertia, with heavy elements located in the extremities of the W 196 R. The water and oil coolers were right at the front, the huge tanks holding petrol and oil in the tail.

In 1955 the front drum brakes were relocated into the wheels on some cars, while three different wheelbase lengths were available: 2150mm (84.6in), 2210mm (87in), and 2350mm (92.5in). The shortest was ideally suited for the tight round-the-houses circuit in Monaco.

The straight eight used desmodromic valve operation, where the valves were closed by the action of the cams and rocker arms, rather than by springs. This allowed the use of larger and heavier valves (50mm (2in) inlet valve

A tubular spaceframe was used that weighed just 36kg (79lb). With the exhausts exiting next to the cabin, good hearing protection must have been needed!

diameter!), and higher rpm. Two valves were used per cylinder. A long tubular inlet plenum chamber was employed with a single throttle body positioned at the front. Long ram-inlets fed air to each cylinder.

The mechanical fuel injection pump, developed together with Bosch and not unlike the ones used in diesel engines, consisted of a casing with eight cylinders which fed the fuel straight into the combustion chambers at a pressure of 100 kilograms per square centimetre (about 1400 psi).

The eight-cylinder in-line configuration was inspired by the famous 18/100 hp Grand Prix car of 1914, in that the cylinders (two groups of four, with central power take-off) were firmly connected to a base plate, through-bolted to an aluminium casing separate from the valve gear housing, and surrounded by a welded-on cooling-water jacket.

Fuel was a highly reactive Esso mixture with code RD 1, concocted from 45 per cent benzene, 25 per cent methanol, 25 per cent 110/130 octane petrol, 3 per cent acetone and 2 per cent nitro-benzene. This blend could eat away an unprotected steel tank overnight ...

The W 196 R's track record was impressive indeed: nine victories and fastest laps, as well as eight pole positions in the twelve Grand Prix races in which it was entered, and Fangio's world champion's titles in 1954 and 1955.

The front suspension used double wishbones and torsion bars. Note how the torsion bar projects through the radiator, and uses a bar within a tube allowing a much longer effective spring. These photos were taken at the Mercedes-Benz Museum in Stuttgart, Germany.

AND THEN THERE'S THE TRANSPORTER!

The legendary Silver Arrows were not the only vehicles that were the talk of the town in the 1950s. The Mercedes-Benz racing department also hit the headlines off the racetracks, with its 'fastest racing car transporter in the world.'

To build the transporter, the X-shaped tubular frame of a production 300 S was extended at the front and rear. A direct-injection SL engine was installed just above the front axle. A synchronised four-speed transmission with involute profiles and a Daimler-Benz single-plate dry clutch were used.

At the front, a double wishbone axle with coil springs and Fichtel & Sachs shock absorbers was installed. At the rear was a swing axle with a hypoid crown wheel and pinion assembly, lowered and separate pivot points for the two axle carriers, coil springs, an additional torsion bar spring and also F & S shock absorbers. At 3050mm (120in), the wheel-base was impressively long.

Hydraulic drum brakes on all four wheels and a pneumatic brake booster from Bosch slowed the vehicle, which had a laden weight of just under 3000kg (6600lb). Safety was further enhanced by a disc brake installed between the propeller shaft and differential, as well as by an exhaust brake – a new feature on a passenger car engine.

The engine itself was very special: in view of the heavy-duty work lying ahead of it, the output of the original three-litre six-cylinder in-line engine from the 300 SL was reduced from 160kW to 143kW (215 to 192hp). With this output and its high torque, this engine easily gave the racing car transporter a top speed of 160-170km/h (about 100-105mph)!

All of these components were hidden beneath a sheet metal skin. The centre section of the driver's cab – from the windshield to the rear edge of the door – was adopted from the 180 production model, only everything was much wider. The front end, set at a steep angle and rounded, flowed into an SL-type radiator grille with a Mercedes star in the centre. The front end's rounded contours were delineated at the sides by rudimentary fenders, unmistakable features of the 180, with integrated headlights and fog lights, and at the front by a sweeping chrome bumper, with large bumper guards, wrapping around the corners of the cab.

Behind the cab mounted in a low position and ahead of the front axle, the transport section of this remarkable vehicle began. The frame, engine, radiator, axles and fuel tank all vanished under complete panelling. The front and rear fenders elegantly extended far towards the rear, enclosing the wheels in an aerodynamically efficient manner. The front and rear fenders were linked by inward-curving panelling flowing into the load platform. The recessed space between the axles accommodated a spare wheel on each side, fixed at an angle and quickly retrievable when required.

The racing car transporter was a sensation on European roads and motorways during its active time from mid-1954 until the fall of 1955, when Daimler-Benz withdrew from racing.

The original transporter was scrapped in 1967 but in 1993, Mercedes had this historically accurate replica built. It's in the Mercedes museum in Stuttgart, where this photo was taken. The transporter is just superb.

The Mini is an oddity: a triumph of technical sophistication applied in a minimalist mode, a car that became a fashion icon but was designed with 'fashion' not even considered, a car that reputedly lost money on every vehicle produced, and finally, a car that was said to set the scene for the revolution in front-wheel drives – but used a gearbox-in-sump (oil pan) driveline approach actually taken by very few subsequent cars.

The Mini – a masterpiece in automotive packaging that has never been beaten. With the gearbox in the engine's sump and the drivetrain turned sideways and driving the font wheels, the engine takes up an extraordinarily small amount of the overall vehicle length. Tiny 10-inch wheels reduce the intrusion of the wheel-wells, and a trailing arm rear suspension with horizontal springs removes any intrusion into the boot (trunk). Between 1959 and 2000, more than 5.3 million Minis were built.

Opposite: (Left) The double wishbone front suspension used rubber in compression as the springing medium, with some other later cars using a sophisticated front/rear interconnected fluid/rubber system. The car bristled with innovation – when Ford pulled one apart soon after it was released, it couldn't work out how BMC was making any money. The answer is that it wasn't!

(Right) The trailing arm rear suspension was amazingly compact – probably unbeaten to this day. This is the view looking downwards with the body removed. This cutaway Mini was photographed at the Haynes International Motor Museum at Sparkford in Somerset, UK.

One reason for the roominess compared with modern cars is the lack of equipment and safety. Doors were thin and used sliding windows, while the absence of airbags and air-conditioning meant the dash could be minimal in size and intrusion. A heater was provided, however.

2000S – TOYOTA PRIUS

The Toyota Prius changed car technology in a way that no other car had done for more than 75 years. If you think that a big statement, consider all the alternative car engines that had been enthusiastically embraced as game-changers – but had then made no lasting impact. The rotary engine: once, nearly every major manufacturer, from GM to Mercedes, had licences to produce the Wankel rotary. But only one car manufacturer persisted – Mazda – and it never became the dominant powerplant in its line-up. Consider the gas turbine: a lot of resources were ploughed into its development – especially in the 1950s and 1960s – but it became mainstream for no manufacturer.

Now consider hybrid cars – those cars that combine a conventional internal combustion engine with an electric motor. Like so many technologies, hybrid drives have been around for 100 years. But until the Prius, no modern car manufacturer sold a mainstream model with hybrid drive. Yet, within a decade of the launch of the Prius, every single car manufacturer in the world was either selling hybrids or working on them for later release.

And it needs to be remembered that Toyota pioneered the hybrid approach at a time when it was being widely ridiculed for doing so so. To (I hope) its everlasting shame, one of the giants of innovation in electric and electronic systems in cars – the German company Bosch – actually issued a press release at the time of the new Prius stating that it could see no future in taking the hybrid path. The vast majority of specialist motoring media poured scorn on the cars in review after review.

Released as a Japanese-only model in late 1997, the NHW10 (ie Series I) Prius introduced the Toyota Hybrid System (THS) which incorporated two electric motors and an internal combustion engine. The electric motor generated 30kW from 940 to 2000rpm and worked with a 288V nickel-metal hydride battery pack (comprising a total of 240 individual D-sized cells). Battery charge was maintained by a regenerative braking system and, where necessary, the internal combustion engine could provide direct charging.

The Toyota Echo-based 1NX-FXE engine in the Prius employed the Atkinson cycle to optimise efficiency. With a swept capacity of 1.5 litres, a 13.5:1 compression ratio, double overhead cams, 16 valves and variable inlet cam timing, the Prius's petrol engine could generate 43kW (58hp) and 102Nm (75ft-lb) – both achieved at 4000rpm. Electric and engine power were combined by means of a special gearbox-like device; the total output from the Series 1 THS was 73kW (98hp).

Weighing 1240kg (2750lb), the Series 1 Prius accelerated from standstill to 100km/h (62mph) in around 16 seconds. However, it should be noted that its real-world on-road performance was not as bad as this figure suggests – the electric power gave instant torque and good flexibility.

Over a conventional contemporary car of equivalent size,

Hybrids are not new, as this photo of the engine bay of a 1914 Galt shows. The Canadian Galt's petrol (gasoline) engine ran a generator that supplied electricity to an electric motor that propelled the car. Extra power produced by the generator was stored in batteries found beneath the floor. The batteries could be used to provide extra power at peak times or to operate the car with the internal combustion engine turned off. This car is one of only two ever built. The photo was taken at the Petersen Automotive Museum, Los Angeles, USA.

exhaust emissions were reduced by about 80 per cent and a fuel consumption by around 50 per cent.

Under the skin of the Prius, the engineering thoroughness was formidable. Everything was superbly made and engineered. The special high voltage connectors, the filtered fan-forced cooling of the battery box, the water cooling of the power converter under the bonnet, and the use of standard Toyota hydraulic transmission fluid in the Power Split Device. The book *The Prius That Shook the World: How Toyota Developed the World's First Mass-Production Hybrid Vehicle* (by Hideshi Itazaki) gives some idea of the thoroughness with which Toyota approached the task, to the extent that, for example, Panasonic had to change the way that it made nickel metal hydride batteries to meet Toyota's durability expectations.

Only 37,000 of the first generation Prius were produced. The second generation, the NHW11, looked much the same as the earlier car, but had significant engineering changes under the skin, including more power. It was sold internationally from 2000, whereupon the car started making a worldwide impact. Since then, the Prius has gone onto a further three generations, and sales have exceeded 4 million.

2010s – TESLA MODEL S

As with hybrids, electric cars are almost as old as the car itself. In fact, an engineering series of books on cars that I have, published in 1919, devotes equal space to internal combustion engines, steam and electric powerplants.

Every major manufacturer produced research electric vehicles, especially in the 1970s, 1980s and 2000s. However, none produced a product that captured the imagination of the car-buying public like Tesla. In an extraordinary development, in 2008 a start-up company released an electric sports car that immediately established the state of the art in saleable electric cars – that car was the Roadster. With the Roadster, Lotus produced the basic car before Tesla equipped it with its

Released in late 1997 for only the Japanese domestic market, the NHW10 Prius was a startlingly innovative car. Before the Prius, no manufacturers were selling hybrid cars; after the Prius, every manufacturer in the world was working on such cars for later release – or already selling them. (Courtesy Toyota)

With its very high performance and eco credentials, the Tesla Model S has made electric cars fashionable. (Courtesy Tesla)

electric drivetrain. The Roadster was the first production car to use lithium-ion battery cells and the first production electric vehicle with a range greater than 320km (200mi) per charge. Only 2450 units were sold, but it set the scene for much bigger things to come.

Tesla released the Model S in 2012. Still in production at the time of writing, the Model S has so far sold 213,000 units – it's the most successful electric car the world has ever seen. The car is available in all-wheel drive and rear-wheel drive models, and with varying battery capacities. Greatest range of any of the versions is about 600km (370mi), with the best performing cars reaching 100km/h (62mph) in just 2.4 seconds in the special 'Ludicrous Mode.'

Irrespective of whether Tesla continues to dominate electric car production or not (and it's likely that the mainstream car companies will soon overtake the small company), Tesla has changed the automotive world.

Acknowledgement: *Some of the text of this chapter is based on material drawn from Mercedes-Benz (Benz Patent Motor Car and Bertha Benz, Mercedes W 196 Grand Prix car and its transporter).*

2015 Tesla Model S P85D chassis, photographed at the Petersen Automotive Museum, Los Angeles, USA. When viewed in this way, the aluminium die-castings and pressed aluminium sheet that make up the body are easy to see.

Chapter 7: Bridges

Getting to the other side

The most wonderful structure of any type that I have ever seen is the Firth of Forth railway bridge in Scotland, UK. I enjoy bridges, and the country that I live in – Australia – has a darned good one in the shape of the Sydney Harbour Bridge. Sailing under it and looking up at its riveted magnificence, I am quite moved. But the Firth of Forth railway bridge is simply in another league.

Why? Well, for starters, the Forth Bridge is just huge. Trains running across it look tiny. Beneath the bridge, height had to be left for tall-masted sailing ships. The bridge is, in round terms, *three times* as long as the Sydney Harbour Bridge. And the Forth Bridge looks so strong. With the failure of the Tay Bridge still fresh in everyone's minds, that appearance of strength is not serendipitous: the public needed to be absolutely certain that here was a structure that could be trusted. Over-engineered? Certainly – but with a purpose. It's perhaps this aspect that makes the bridge like no other that will ever be built, it not only had to be strong but also *look* strong.

In contrast, the Brooklyn Bridge in New York appears almost under-engineered, with its spidery web of cables supporting the deck. The Brooklyn Bridge was started in 1870 and finished in 1883 – so construction of the Forth

The Brooklyn Bridge, against the famous New York City skyline. Built in 1883, it still carries about 150,000 vehicles each day.

Bridge began as the Brooklyn Bridge was being finished. But despite being from a similar era, the engineering could not be more different. Described by most older references as a suspension bridge – but now more correctly called a cable stayed/suspension bridge – the Brooklyn Bridge was extraordinarily daring in its design. And, unlike the Forth Bridge, the Brooklyn is accessible to pedestrians. When we walked across it, two things struck me: first, the complete obliviousness of the thousands of selfie-photographing pedestrians to the magnificence of the structure that was supporting them, and the exquisite contrast between the huge stone towers and the finery of supporting cables.

A quite different bridge is the Iron Bridge in Shropshire, England. A century older than the Brooklyn Bridge, this extraordinary structure was the first major bridge in the world to be made of cast iron. We visited it early on a winter's morning, the bridge and the eponymous surrounding town shrouded in light fog. A relatively small structure compared with the giant bridges so far described, it moved me in a quite unexpected way. I thought it would be of interest, certainly, but it was only when I saw the woodworking style of joints used to hold much of the structure together that I realised I was seeing the results of an engineer using proven engineering approaches – but with a material being used in a brand new way. The bridge is open to pedestrians: it's one to walk across, marvelling that nearly 250 years after being built, a bridge engineered on the very edge of available technology is still standing.

Moving bridges add yet another dimension. My favourite is Tower Bridge in London. Despite its old-world appearance, Tower Bridge is a contemporary of the Forth Bridge and, in fact, uses steel supporting towers. (They were clad in stone to provide a better aesthetic match for surrounding buildings.) A combined bascule and suspension bridge, the sophistication of the technology becomes visible only when you descend beneath the bridge to the engine room, and see the huge stationary steam equipment that allows the bridge to open and close.

Not quite a bridge, but at home in this chapter, is a modern structure – the Falkirk Wheel in Scotland. Opened in 2002, this extraordinary construction is part sculpture, part boat lift, part – well, I am not quite sure. The lift transfers canal boats from one level to another, and was built as part of the reinvigoration of the canal system in Scotland. Used for tourist and residential canal boats, it's a magical extravagance that's delightfully impressive when seen in the metal.

As its name suggests, Iron Bridge is made from cast iron – the first major bridge in the world to use this material. The bridge is nearly 250 years old.

The Forth Bridge uses three huge symmetrical cantilevers, joined by trusses. This design approach gave the required clear spans and also allowed the bridge to be self-supporting during construction.

THE FORTH BRIDGE

A mighty structure

In the 1870s it was easy to take the east-coast train from London to the north of Scotland. But beyond Edinburgh came two interruptions to the smooth and rapid journey. First was the mile-wide Firth of Forth estuary, then a little further north, the even wider Firth of Tay. Passengers needed to disembark from the train and catch ferries across these stretches of water. As well as being inconvenient, this also increased journey times. To take advantage of this, west coast railway companies were threatening to swing their lines eastwards, allowing rail passengers to bypass both barriers. And so plans were made to first bridge the Tay, and then the Forth.

Thomas Bouch was the engineer selected to design the Tay Bridge. Bouch had designed roll-on/roll-off ferries that allowed railway wagons (although not passenger carriages) to be easily transported, and these 'floating railways' were seen to be a wonder of the age. Bouch also had extensive experience in constructing railways and smaller bridges, and so when he put forward plans to bridge both the Tay and Forth estuaries, the directors of the North British Railway Company took him seriously. The Tay Bridge was first.

The Tay Bridge used a spidery elegance of lattice girders constructed from cast and wrought iron, resting on iron piers. It comprised 85 spans, with 72 of the spans supporting girders below the track, and the remaining 13 spans above the track level. These 'high girders,' as they were known, gave navigational clearance of 26.8m (88ft) to the high-water mark. The bridge was completed in early 1878; at over 3km (2mi), it was then the longest bridge in the world.

The bridge was a marvel of Victorian engineering. Bouch was awarded a knighthood, and the foundation stone for his next bridge design – the Firth of Forth – was laid. But then, at about 7.15pm on the stormy night of 28 December 1879, the high girders of the Tay Bridge collapsed, taking with them a train, six carriages and 75 people into the icy waters. Public confidence in Thomas Bouch also collapsed, and Parliament quickly passed an Abandonment Act that stopped further work on Bouch's Firth of Forth bridge design.

An inquiry into the Tay Bridge disaster held Sir Thomas Bouch chiefly to blame for the collapse – he had not made an adequate allowance for wind loadings in the design. That such a simple mistake could be made by an eminent engineer threw into doubt the ability not only of Thomas Bouch, but of large scale engineering in general.

The Firth of Forth was too deep an estuary to use a simple girder-and-pier design like that used over the Tay. In fact, Bouch's proposed bridge was a suspension/cable stayed design with two large spans, the centre towers supported on Inchgarvie (an island fortuitously located in about the middle of the estuary). Rather than having many spans of 75m (245ft) like the Tay Bridge, the Forth Bridge would need two spans of over 500m (1600ft) each! This was a huge challenge indeed – especially in the environment of public doubt. After all, who would want to catch a train that passed over a bridge that could collapse just as the Tay Bridge recently had?

Multiple proposals were put forward for the Firth of Forth bridge, with most being variations on suspension designs. However, the design suggested by Benjamin Baker and John Fowler was nothing like a suspension bridge. Instead, it was based on three assemblies, each with two enormous cantilever arms extending outwards either side of vertical towers. The three cantilever structures were each to be about 457m (1500ft) long and be joined by trusses 107m (350ft) in length. The height to the top of the proposed bridge was 110m (361ft), and the railway track was to be 46m (150ft) above high water level. The total bridge length was to be 2467m (8094ft). Wind loadings used in the calculations were 56lb/ft^2 – Bouch had used a figure of just 10lb/ft^2 for his Tay Bridge. Compared with the Tay Bridge, the proposed Forth Bridge looked strong enough to cope with anything – and that, of course, was one requirement of the design.

Almost three years to the day after the Tay Bridge disaster, on December 21, 1882, the contract for the construction of the Forth Bridge was awarded to Tancred, Arrol & Co. William Arrol, the contractor, was, like Benjamin Baker, a man of exceptional ability. He went to work in a cotton mill at aged nine, and at 14 became a blacksmith's apprentice. In 1863, he joined a building and bridge construction company, and then in 1868 started his own company. He was the successful tenderer for the aborted Bouch Forth Bridge design, and went on to become the successful tenderer for the Baker design. Designer Benjamin Baker was also a man with practical hands-on experience. He completed an apprenticeship at a South Wales Ironworks, and then moved to London where he worked on the construction of Victoria Station. As an engineer, he subsequently worked on projects as diverse as London's underground tube train network and the Aswan Dam.

The bridge is made from steel (a relatively new material at the time), and is riveted together. The bridge was assembled twice, with each small section initially erected on land.

Construction of the bridge began with the sinking of caissons and the building of piers. This took place in the period 1882-1886, with the next four years devoted to building the bridge superstructure. The quantities of materials used were enormous: 54,000 tons of steel; 194,000 cubic yards of granite, stone and concrete; and almost 7 million rivets. No falsework was used: the bridge supported itself at every stage of construction. Readers of the *The Scotsman* newspaper in March 1888 learned:

"The workers today are practically standing upon their labours of yesterday. As soon as a fresh round of steel plates is added to the tubes, or an additional girder section riveted to the top arms, the platforms with their freight of men and cranes and other mechanical appliances are slid out correspondingly and a new piece of work is begun which again, when completed, will give the necessary standing support for a further extension. Every piece of work done becomes the basis of another advance ..."

At the height of construction, more than 4000 men were employed; that 57 of them died was seen at the time as being a remarkably low number. The bridge was completed in December 1889. Both Baker and Arrol were knighted for their work.

The bridge has stood proud and strong – very strong – ever since. About 200 trains a day cross it, tiny and model-like in appearance against the colossal structure. The bridge dominates the landscape, the immensity of its engineering shrinking the villages on both shorelines to visual insignificance.

The Forth Bridge is commemorated by UNESCO as a World Heritage structure. This body states:

"This enormous structure, with its distinctive industrial

aesthetic and striking red colour, was conceived and built using advanced civil engineering design principles and construction methods. Innovative in design, materials, and scale, the Forth Bridge is an extraordinary and impressive milestone in bridge design and construction during the period when railways came to dominate long-distance land travel. The Forth Bridge is a masterpiece of creative genius because of its distinctive industrial aesthetic, which is the result of a forthright, unadorned display of its massive, functional structural elements."

The approach viaducts to the Forth Bridge are huge and impressive in their own right, with multiple spans of 50m (165ft) and a height of 46m (150ft).

THE ERASMUS BRIDGE– Two bridges in one

The Erasmus Bridge was completed in 1996 and crosses the Maas River in Rotterdam, the Netherlands. The bridge combines a cable-stayed span with a bascule (opening) section. The main span is 284m (932ft) long and the bascule is 55m (180ft) long. Including approaches, total bridge length is 548m (1800ft).

The Erasmus Bridge's claim to fame is its extraordinary elegance. Designed not by an engineer but by the architect Ben van Berkel, the bridge uses a 139m (456ft) tall pylon that instead of being straight as is normal for a cable-stayed bridge, has a distinctive kink. Interestingly, the bent pylon, although more complex to design, could be made shorter than an equivalent straight pylon, saving materials.

The 1800 tonne pylon is constructed from welded steel plate, 50mm (2in) thick in high-load areas and 20mm (¾in) thick elsewhere. The front cables comprise 32 stays, each made up of between 30 and 48 galvanised and polyethylene-coated strands, all hidden inside a high-density polyethylene cover.

The slender deck of the main span provides 12.5m (41ft) clearance for shipping. Rather than the deck being covered in bitumen or concrete, an impermeable mastic epoxy layer is used to save weight. Tall vessels pass through the open bascule part of the bridge. One of the largest of its kind in Europe, the moveable deck weighs 1560t and is 52.3 x 35.4m (175 x 116ft). Opening and closing times are 120 and 135 seconds respectively.

At night the bridge is simply stunning.

Using a unique kinked pylon, the cable-stayed Erasmus Bridge in Rotterdam is a wonderfully elegant design. Not visible here is an opening section, positioned to the right.

THE GOLDEN GATE BRIDGE

One of the most elegant bridges in the world

The idea for spanning the Golden Gate in San Francisco, USA, had been considered for many years, but it was James Wilkins, a structural engineer and newspaper editor, who begun strongly promoting the project. This was brought to the attention of San Francisco City Engineer Michael O'Shaughnessy, who became interested, and began to consult a number of engineers across the United States regarding the feasibility and cost of building a bridge.

The first serious proposal for the bridge was presented in 1921, when a Chicago-based engineer, Joseph Strauss, developed a bridge design that featured an odd mixture of cantilevers and suspension approaches. It took O'Shaughnessy 18 months to release the cantilever-suspension design to the public, where it was labelled ugly by the local press.

Raising political support for the bridge was difficult, and financial support even more so. Finally, in 1928, the Golden Gate Bridge Company was formed, and in 1929, long after he first become involved, Strauss was appointed Chief Engineer. Also in 1929, Leon Moisseiff, O H Ammann, and Charles Derleth were appointed to serve as an Advisory Board of Engineers, working alongside Chief Engineer Strauss. Strauss also appointed Charles Ellis to work with this board, acting as its Secretary.

The use of a suspension bridge – rather than the original ugly Strauss design – was first nominated by Leon Moisseiff; however, the timing of the change in approach is not known. A report written in 1937 by Strauss, simply states, "... In the interval which had elapsed, any advantages possessed by the cantilever-suspension type bridge had practically disappeared and on recommendation of the Chief Engineer, the cantilever-suspension type was abandoned in favour of the simple suspension type."

Charles Ellis was responsible for directing all the detailed stress calculations, as well as the development of the specifications, contracts and proposal forms. However, for reasons which are not known, Strauss fired Ellis about 12 months before construction started, so robbing Ellis of much of the credit for the bridge design.

Rather like the Hoover Dam (see Chapter 4), a specialist was asked to improve the styling of the bridge. Irving Morrow, an obscure San Francisco-based architect who designed houses, joined the project in April 1930. Told by Strauss to "make it beautiful," Morrow simplified the pedestrian railings, redesigned the light posts, and added wide, vertical ribbing to the horizontal tower bracing. Morrow also designed a lighting system to

The Golden Gate Bridge's grace disguises its sheer size. The main span is 1280m (4200ft) long and the towers are 227m (746ft) high. Clearance under the bridge is 67m (220ft).

The art deco styling of the towers was the work of architect Irving Morrow, who was told by engineer Joseph Strauss to "make it beautiful."

The main cables are 92.4cm (36.4in) in diameter and 2332m (7650ft) long. The total length of galvanized steel wire used in both main cables is 129,000 km (80,000mi). The deck at the middle of the bridge is designed to move vertically by as much as 5m (16ft).

illuminate the bridge, but it wasn't until the bridge's 50th anniversary that these lights were installed.

One of the bridge's most distinct characteristics – its colour – was also the responsibility of Morrow. As supplied by the steel makers, the steel was coated in red lead primer. Watching the towers rise in this colour inspired Morrow to think of using a similar colour on the final bridge design. He consulted widely with other architects, engineers, painters and sculptors, also considering the use of black, grey and silver. Italian-American sculptor Beniamino Benvenuto Bufano submitted these comments to Morrow:

"I have been watching very closely the progress of the towers on the Golden Gate Bridge in its structural beauty, its engineering and architectural simplicity – and, of course, its colour that moves and moulds itself into the great beauty and contours of the hill – let me hope that the colour will remain the red terracotta because it adds to the structural grace and because it adds to the great beauty and the colourful symphony of the hills – and it is because of this structural simplicity that carries to you my message of admiration."

The final colour chosen is dubbed Golden Gate Bridge International Orange. It comprises Cyan (0%), Magenta (69%), Yellow (100%) and Black (6%).

The construction of the bridge was made difficult by the tremendous currents of the strait. Partly to avoid these, Strauss made the decision that the pier for the north tower would be placed on a stone outcropping off Marin County shore. The south pier, built in the water, required that a massive sea wall be first constructed around the site. Divers could work for only four 20-minute shifts between tides.

Once the piers were constructed, the towers went up smoothly. The piers are 12m (40ft) high, and the towers on top of them rise another 210m (690ft). The towers contain 22,200 tons of steel and use 600,000 rivets. The columns of each tower are 27m (90ft) apart, in line with the main supporting cables. During construction, an earthquake struck. Workers, perched atop the incomplete towers, said the towers swayed back and forth over 5m (16ft).

The cables are 2332m (7650ft) long and are made from 27,572 solid wires of 0.5mm (0.196in) diameter with 80,000psi yield strength. The wires are formed into 61 strands that were squeezed by a radial compression machine into one round 924mm (36.4in) cable, banded every 15.2m (50ft) by clamps of cast steel. Once the main cables were strung, the vertical suspenders were installed. These use 150 solid wires of 0.5mm (0.196in), again with 80,000psi yield. The wires are bundled to give cables with a diameter of 69mm (2.7in). With the supporting cables in place, the roadway could then be constructed, extending out from each tower and with a camber (vertical rise) of 3.2m (10.6ft).

The bridge was opened on 27 May 1937. One account says that 200,000 people skated, pogo-sticked, or walked across the bridge in celebration. Subsequently, the bridge deck has been strengthened to reduce deck movement in high winds, and to improve the bridge's resistance to earthquakes.

THE FALKIRK WHEEL

A canal boat-lift like no other

One of the most interesting engineering structures you can see in Britain is the Falkirk Wheel. To many of us living in other parts of the world, its function is odd: it's designed to transport canal boats from one canal to another. But its shape is also weird: with its huge curved axe-like arms, it's like nothing else that you'll find. And then there's the mechanism – massive, exposed cogs that obviously perform an engineering function – but what?

So let's take a look at a marvel that mixes artistry, engineering and function.

The canals

Before you can understand the function that the Falkirk Wheel performs, you need to know about the canals it links. The Forth & Clyde canal was constructed in the period 1768-1773, and fully opened in 1790. The 56km (35mi) long canal connected Grangemouth (on the Firth of Forth) to Bowling (on the Firth of Clyde). It therefore connected the east and west coasts of Scotland. In

The Falkirk Wheel is a canal lift that moves canal boats from one level to another. The aqueduct at top-right shows the upper level, and the water below shows the lower level.

The Falkirk Wheel in operation. Just visible in the round opening is the end of a canal boat that, complete with the water in which it is floating, is being moved from one level to the other.

The huge timing cogs keep the gondolas level as the wheel rotates. The wheel is driven by ten 7kW (9hp) hydraulic motors working on the central axle.

addition, the canal had a short spur to Glasgow. The canal boasted no less than 40 locks, 25 aqueducts and 32 opening or swing bridges. It closed to navigation in 1963 when a highway was routed across the canal at low level.

The Forth & Clyde canal might have needed to climb mountains, but the Union Canal was quite different. For most of its length it was level – running along a single contour. This canal connected Edinburgh to Falkirk. It was 50km (31 miles) long and was constructed between 1818 and 1822. It closed in 1965 – again, the final nail being the routing of a highway across it.

At Falkirk the difference in level between the two canals was 34m (110ft). Canal vessels were able to move from one canal to the other via a flight of eleven(!) locks. To do this took most of the day and used 3500 tonnes of water.

In their day, the canals were a major transport link, carrying up to 200,000 passengers per year. The boats were drawn along the canal by horses walking alongside. But not all vessels were small: the Forth & Clyde canal was the world's first sea-to-sea ship canal and had to accommodate tall-masted vessels. This meant the bridges over the canals had to open or swing to allow the ships passage.

For decades after closing, the canals decayed. Locks were filled-in, bridges crumbled and the canals were maintained only for safety and to act as drains. Then, in the mid-1990s, British Waterways – the canal owners – decided to revitalise the full length of the two canals. Public interest and support were high, and government made funding available. Locks were restored, new sections of canal dug to replace those covered by urban development, rubbish was removed and bridges built.

But what was going to happen at Falkirk, where the canals differed in height by 34m (110ft) – and the original locks, that had not been used since 1933, were long buried?

The wheel

To rejoin the canals, an approach had to be decided upon. Building new locks was not considered a viable option, so the device that would transport boats from one canal to the other needed to be mechanical. All sorts of options were considered, including an overhead mono-rail that transported basins of water containing boats; a vast spoon-shaped see-saw; and even a vertical lift concealed within a cylindrical waterfall.

However, over the 5-year design process, the concept of a turning wheel was decided upon. In its initial stages, this looked rather like a huge Ferris wheel containing two basins for holding boats and water. By 1999 the design used four hanging gondolas. However, even with the construction company by then chosen, British Waterways had a change of heart. In what must have been an intense month of meetings, 20 engineers, architects and steel fabricators devised the final design.

And what was that design? Opened in 2002, the Falkirk wheel is 35m (115ft) high and 27m (88.5ft) long. It comprises two enormous steel arms, for aesthetic reasons shaped into axe-like curves at each end. The arms support two gondolas, each capable of holding 250t of water and boats. The arms were fabricated off-site in a factory, assembled in that factory, and then disassembled for transport to the site. There they were bolted (not welded) together, with over 15,000 bolts used. The steel assembly has a mass of 1200t.

The 23.7m (77.8ft) long gondolas, each of which has a mass of 50t empty, ride on wheels mounted within a circular opening at each end of the arms. To ensure that the gondolas remain level as the Wheel rotates, each gondola is geared by means of large (very large!) timing cogs to the central part of the axle. This gearing is passive – it needs no extra motors, as the motive force is provided by the turning of the wheel as a whole. The wheel itself is driven by ten 7kW (9hp) hydraulic motors working on the central axle.

Because each gondola has the same mass, irrespective of how many boats it contains, the wheel remains balanced. (Each boat displaces its own weight in water – so it doesn't matter if the gondola is full of boats or is empty ... it weighs the same.) It takes only 1.5kWh to turn the wheel through a boat transfer operation. To retain balance, the water levels must be the same in both gondolas. In fact, computer control of the canal levels keeps the levels in the gondolas within 75mm (3in) of each other. Steel gates are used at each end of the gondolas, and rubber seals shut off the 50mm (2in) gap between the end of the gondola and the canal.

There were significant design challenges in developing the wheel. There are 600t, 100 per cent stress reversals during the tuning of the wheel, and the design had to cope with potential boat impacts (quantified as a 40t load applied at aqueduct level) and a wide temperature range.

The Wheel is located at the end of a 2km (1.2mi) extension to the Union Canal. This extension runs through a new tunnel, and then across a 104m (340ft) long aqueduct that leads directly to the wheel's upper gondola. To aesthetically match the profile of the Wheel, the aqueduct had to appear to be floating. This resulted in very high loads being transferred through only small areas of reinforced concrete. The upper parts of the aqueduct hoops are actually not concrete, but are instead reinforced plastic.

Seeing the Falkirk Wheel in the metal, you're torn – is this a whimsical exercise in silliness, or an energy-effective engineering structure? The answer is that perhaps it is both – artistry integrated with engineering.

THE OTOWI SUSPENSION BRIDGE – Concrete, steel and timber: the bridge over the Rio Grande

The Otowi Suspension Bridge, built in 1924, spans the Rio Grande river in Santa Fe County, New Mexico, USA. The single span structure is 53m (174ft) long, with a wooden deck 3m (10ft) wide. The two supporting towers are made from reinforced concrete, and are just over 6.4m (21ft) high. The supporting cables, 86.3m (283ft) from anchor point to anchor point, are 57mm (2¼in) thick.

The bridge is on the National Register of Historic Places, and its Statement of Significance says: "The Otowi Suspension Bridge offers a striking appearance as it spans the Rio Grande. Its scale and proportions blend harmoniously with the low hills sloping down to the crossing site. Its combination of wood, concrete and steel materials creates an aesthetic uncommon in other bridges."

And the bridge is significant for another reason. All the people and materials that travelled to Los Alamos, the birthplace of the atomic bomb, passed across this bridge.

A bridge that looks nothing like its near-100 year age. The bridge is also significant in its contribution to the end of World War II, and the subsequent Cold War.

THE TOWER BRIDGE – Unique in appeaance and with much of its engineering hidden

Opened in 1894, Tower Bridge in London, UK, is a unique design.

Built by Forth Bridge contractor William Arrol and Co, Tower Bridge uses two suspension spans, two trusses and two bascules (opening sections). The steel framed, stone-clad towers rise 65m (213ft) and are spaced (230ft) apart. Joining them high above the Thames River are two trusses that contain pedestrian walkways. The trusses also conceal tie-bars that resist the pull of the suspension spans either side of the towers.

Beneath the walkways lie the two bascules, each about 30m (100ft) long. Each bascule weighs 1200t and is balanced delicately on pivots. The pivots – 533mm (21in) in diameter, 14.6m (48ft) long and weighing 25t – are carried in roller bearings. Each bascule counterweight comprises 290t of lead and 60t of iron. The original opening machinery used a steam and hydraulic system. Two stationary steam engines, each developing 270kW (360hp) pressurised water to 55bar (800psi). This water was stored in hydraulic accumulators, ready for use. When the bridge needed to open, a process taking about 5 minutes, a hydraulic motor rotated pinions that meshed with curved racks (quadrants) attached to each bascule. (In 1974, the steam-hydraulic system was replaced with a new electro-hydraulic system.)

Either side of the opening section are two suspension spans, that in fact are longer than the combined bascules. Each suspension span is 82m (270ft) long and uses 'cables' that actually comprise steel plates 25mm (1in) thick. Unusually for a suspension bridge, the height of the cables differs at each end, giving a better balanced appearance.

The bascules are still raised about 1000 times a year, and the bridge carries over 40,000 people each day.

The original operating machinery is preserved and can be inspected. It comprises two stationary steam engines that pressurised water to 55Bar (800psi). This in turn operated hydraulic motors that rotated pinions bearing on racks attached to the opening bascules.

Tower Bridge in London comprises suspension, truss and bascule designs. The bascules use heavy counterweights hidden beneath the piers.

Chapter 8:

Going fast

On land and in the air

I've been fascinated for a long time by vehicles that go fast, but I think it was Cyril Posthumous' book, *Land Speed Record: A Complete History of the Record-Breaking Cars from 39 to 600+mph*, that really ignited my interest. It was probably the first edition, published in 1971, that I was reading in the mid-Eighties. I don't know if it was his intention, but Posthumous infused in his writing an almost mystic heroism in the drivers – and their machines.

From the elegant, tiny Stutz Black Hawk of Frank Lockhart (the first car in the world to use intercooling) to the massive vehicles driven by Malcom Campbell; and from the tragic demise of John Parry-Thomas in his drolly-named 'Babs' on Pendine Sands, to Craig Breedlove's Spirit of America, the first of a new wave of very successful Land Speed Record cars built in almost backyard conditions and featuring jet thrust as the motive power – none of these cars advanced the technology of normal road-going cars, despite numerous contemporary protestations to the contrary. The Land Speed Record machines were uniquely designed and developed for just one purpose – going fast over the flying mile – and were quite pointless in other environments. Simply, they were built for the hell of it – to go faster than anyone had before.

But many of the other machines covered in this chapter go fast not as the primary purpose, but as part of achieving broader goals. The 1930s Pioneer Zephyr, the world's first stainless-steel diesel train, could set both high top speeds and maintain high average speeds. But it also had low running costs, and cutting-edge

The 1950s Comet was the first passenger jet aircraft – it flew higher and faster than any other. The aircraft design was a brave step into the unknown – with fatal consequences. Three Comets were catastrophically lost through in-flight break-ups. This photo shows part of the recovered roof panel surrounding the square automatic direction finder window, from which the original fatigue cracks spread on one of the lost aircraft. Going faster than anyone else poses enormous risks. (Photographed at the Science Museum, London)

interior comfort and decoration, requirements in the USA to lure people away from their cars and back to railways. The fast steam locomotives of the same era in Britain were mostly used on the passenger routes to the north, two competing railway companies wanting to lower trip times on the run to Scotland. But the locomotives had to work on existing tracks, using existing signalling and rolling stock. Perhaps indicative of the fact that these locomotives weren't designed just to go fast was that Mallard, still the steam record holder to this day, sustained damage when setting the record.

The world's fastest-ever aircraft, the X-15, was developed as the Cold War intensified (see Chapter 3). Going fast just for the sake of it was not the point: developing aircraft that could go faster – and higher – than the opposition was certainly strategically important.

So this chapter covers machines that vary from what could almost be seen as quirky – through to those built with the deadly serious intent of improving a country's defence. But in every case, the designers and engineers dared to imagine an outcome that had never before been achieved. And in most cases, achieving that outcome required taking real risk – financial, reputational, and to human life.

The design or development of every machine shown here is brave in a very special way.

THE 1000HP SUNBEAM

The first car in the world to travel at over 200mph

It's the 29th of March, 1927. The location is the long stretch of sand at Daytona Beach, Florida. The car, the like of which has never been seen before, is an ultra-streamlined machine of imposing length – all 6 metres (20ft) of it. Under the skin there are two enormous engines, each a V12 of over 22 litres (1343 cubic inches) capacity. Power? Nearly 750kW (1000hp). The man behind the large, thin-rimmed steering wheel is Major H O D Segrave.

And he is about to go faster than anyone has ever driven a car ...

The Sunbeam was a radical departure from previous Land Speed Record cars. Its all-enveloping, low-drag body was so different that it was given the unkind nickname of 'The Slug' – but it was a slug that topped 200mph (320km/h). Here it is pictured at the British Motor Museum at Gaydon, UK.

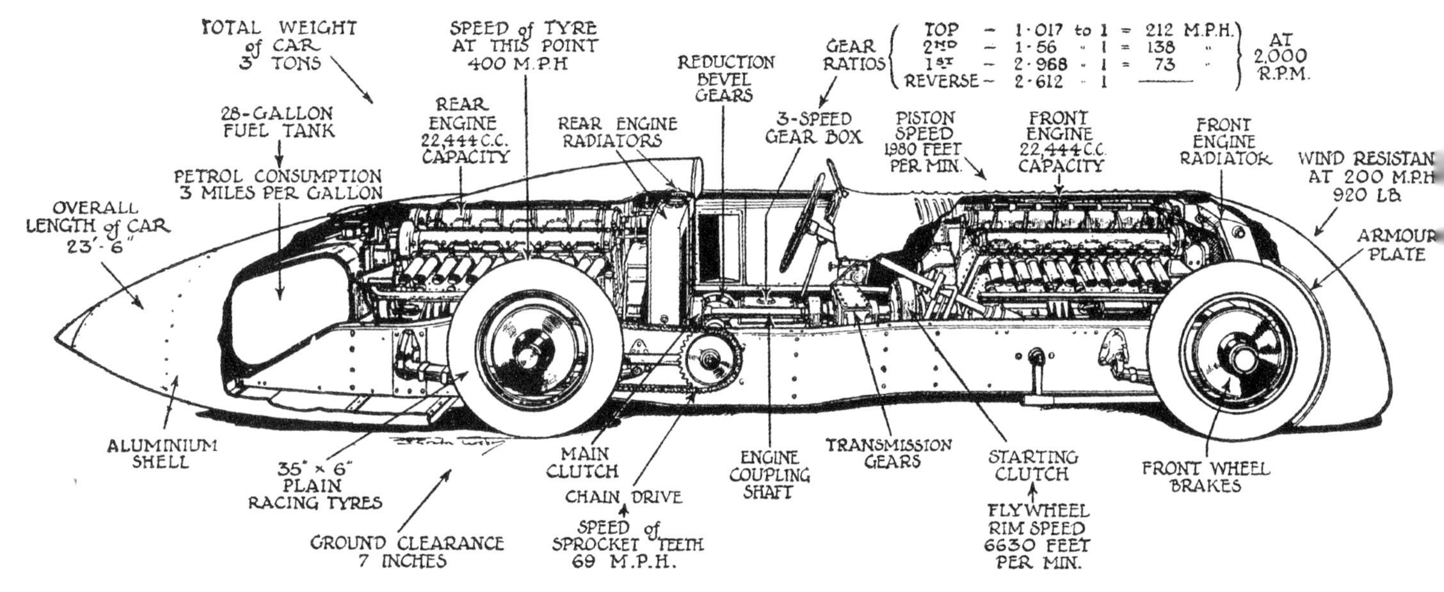

This contemporary drawing clearly shows the mechanical make-up of the car. Each of the two 22-litre engines was enormous.

The 1000hp Sunbeam – otherwise known as 'Mystery' – was designed by Louis Coatalen, chief engineer of Sunbeam, a major car manufacturer in the United Kingdom. Detail design work and construction were left to Captain J S Irving – a name that was to be well associated with later Land Speed Record cars.

An enormous single tachometer and duplicated instruments for the two engines, all behind a thin, wood-rimmed steering wheel.

Two 22-litre engines

The new record contender aimed to achieve success through two main technical approaches – a lot of power achieved through the use of two engines, and low aerodynamic drag achieved through all-enveloping streamlining.

The engines comprised Sunbeam Matabele aircraft engines. The 12 cylinders were cast in blocks of three, and arranged in a V angle with an included angle of 60 degrees. Bore was 122mm (4.8in) and stroke 160mm (6.3in). Each cylinder used four valves, operated by overhead camshafts, and each cylinder used two spark plugs. Four six-cylinder magnetos were used on each engine to provide ignition. Carburetion was by two Claude-Hobson carburettors positioned within the vee. Power output was 373kW (500hp) at just 2000 rpm. The rear engine straddled the rear axle, while the front engine was placed just behind the front axle. Each engine had its own radiator positioned ahead of it, and each engine was mounted in a subframe that, in turn, was mounted to the main chassis by three ball joints.

The engines were mounted back to back, with the small flywheels pointing towards the driver. A system of clutches allowed the front engine to be used to start the rear engine. To reduce the size of the frictional clutches and gearbox, engine speed was increased by 2.5 times to lower the required torque handling. Following the 3-speed gearbox, speed was reduced by further gearing before final drive was made to the rear wheels by two chains. During development, the car was placed on a custom axle dynamometer that used Heenan and Froude retarders. Contemporary accounts say that over 1000hp was measured by the dyno.

Mechanical drum brakes

The chassis of the car used two longitudinal deep-channel girders, braced with channel section cross-members. The longitudinal chassis members were 355mm (14in) deep, spaced 1.08m (42½in) apart in the middle of the car, but tapering (when viewed from above) at each end. Given the large external dimensions, the wheelbase and track of the car were relatively small at 3.58m (11 feet 9 inches) and 1.57m (5 feet 2 inches), respectively. Half-elliptic springs and solid axles were used front and back, with the rear axle kinked downwards in its middle to avoid the engine above. Four-wheel mechanical Dewandre drum brakes were fitted, and four Hartford friction dampers were used per axle.

The shape of the car was extraordinarily unusual. The top of the bodywork was only 1170mm (46in) above the ground. The air intake for the front radiator was via a rectangular opening let into the otherwise curved front. Openings in the side of the bodywork provided air exits for the front radiator, while scoop inlets were used for the rear radiator and engine. The body shape was developed through the use of models tested in the Vickers aircraft wind tunnel at Weybridge. These tests indicated that on the full-size car, aerodynamic downforce at the front could reach 320kg (700lb), while there would be 90kg (200lb) of lift at the rear. (However, other sources suggest that the rear diffuser was designed to provide rear downforce – so these figures may have been reversed over time.) The body was made of aluminium. Complete with fluids, the car weighed 4 tons. At the time, the car's shape led to a rather irreverent nickname – 'The Slug.'

30,000 watched the record being broken

When the car reached Daytona, it had never been driven at high speed – there was no suitable test surface available in the UK. A crowd of 30,000 lined the dunes on Tuesday, 29 March, 1927. At 9.30am, Segrave started his first run, driving north into the wind. As with all Land Speed record-breakers of the time, Segreve changed up the gears as he went – at 145km/h (90mph) changing into second, and at 217kmh (135mph) into third. Side winds pushed the car about and he knocked over some marker flags; the car was slow to respond to steering input. Reaching the end of the course, Segrave braked – but little happened. To slow sufficiently, he had to steer into the sea, the shallow water achieving what the brakes had not. On the return run, the wind again blew the car about, but he was able to hold it on course.

The timing officials checked their calculations, and then announced a new Land Speed Record: over the flying two-way mile, a speed of 327.97km/h (203.79mph). It was the first time anyone had exceeded 200mph (322 km/h). However, Segrave's record stood for only 11 months before it was beaten by Malcolm Campbell, again on Daytona Beach.

Note the long, aerodynamically sophisticated tail that includes a rear diffuser. The car was wind-tunnel developed.

WORLD'S FASTEST HELICOPTER – Upgraded engines, rotors and streamlining

In 1986, a modified Westland Lynx set the world helicopter speed record in England.

Special rotor blades were fitted that used composite construction, and comprised four different aerofoil shapes along their length. Inboard, a high-lift design was used where the effective airspeed was lower, while towards the tips, a near-sonic section was used to prevent blade stall or compressibility effects.

To provide the required power, uprated 1200shp (shaft horsepower) Rolls-Royce Gem 60 engines were fitted. These engines used a water-methanol injection system and were cleared to use 'emergency power' for five minutes at a time. An uprated main gearbox from a Westland 30 helicopter was fitted. A Westland 30 also contributed a new low-set tailplane with vertical fins. Drag reduction measures included removing the windscreen wipers, aerials and steps.

On the calm evening of August 11th, 1986, with Trevor Egginton at the controls and flight engineer Derek Clews in the left-hand seat, the helicopter flew low-level over the measured 15km (9.3mi) course, across the Somerset levels between East Huntspil and Hartlake Bridge. A convenient canal plus cameras and observers under the flight path ensured the Lynx stayed on a straight-line course.

Following two practice runs, one in each direction over the route, the crew went for the record, officially achieving a best speed of 412.93km/h (256.58mph) on one run and an average speed of 400.87km/h (209.10mph). In doing so it became the world's fastest helicopter, a record it still holds.

The modified Westland Lynx helicopter, holder of the world record at 412.93km/h (256.58mph). It was photographed at the Helicopter Museum, Weston-super-Mare, UK.

SPEED AND BEAUTY – THE PIONEER ZEPHYR

Diesel power, high speed and stainless-steel construction

The debut in 1934 of the Pioneer Zephyr was one of the turning points in railway travel. For underneath the Zephyr's brilliant silver exterior lay a revolutionary two-cycle diesel engine that would soon replace the traditional steam engine, changing forever the nature of the entire railroading industry. The 97½-ton train itself was an impressive 60 metres (196ft) of fluted, stainless steel, a lightweight material never before used in the construction of railcars. Unlike most contemporary steam locomotives, the Zephyr was also fully streamlined; this together with the diesel power, allowing a maximum speed of 180km/h (110mph) – and, unofficially, even faster. Inside the train, the Zephyr provided luxury accommodation today's rail travellers can only dream about.

The Pioneer Zephyr – a masterpiece in streamlined stainless steel, featuring a diesel engine. It could reach 180km/h (110mph)) and maintain high average speeds. Photographed at the Museum of Science and Industry, Chicago, USA.

The Pioneer Zephyr was the first to use the Winton 8-201A 450kW (600hp), two-cycle diesel engine. It gave good performance with low operating costs.

Below: The train was constructed using fluted stainless steel sheet, spot welded together. This construction process had never before been used.

Getting passengers back to rail

In 1934, the year the Zephyr made its first run, some 50,000 steam locomotives were puffing their way back and forth across the United States, racking up 17.8-billion passenger miles and 268-billion ton-miles of rail transportation. By 1961, following the Pioneer Zephyr's lead, the country had become completely 'dieselised.' Approximately 28,500 diesel units produced 20-billion passenger miles and 565-billion ton-miles. In other words, the diesel-powered units were capable of doing almost four times the work performed by their predecessors, the steam locomotives. But in the early days, the advantages of the diesel engine were not so evident – especially when rail passenger traffic was in free-fall. In 1924, the Burlington Railroad carried a whopping 18 million passengers. Only five years later, as the automobile's popularity increased, that figure dropped to 13.8 million. And by 1933, in the midst of the Great Depression, only 7 million travellers rode the rails.

Luckily for the railroad industry and the travelling public, a figure appeared on the scene in the early 1930s, with the know-how and determination to build a train to meet the demands of the 20th century traveller. His name was Ralph Budd, and he became president of the Burlington Railroad in 1932.

By the time Budd arrived at the Burlington, the private car had edged out trains as the favoured means of transportation, even for long distance travel. Budd was among the first to admit that "the loss of railway passenger traffic during the last decade has been caused by a shifting from the railways to the highway, and not from a decline in total travel. In fact, the total passenger one-mile units of travel have greatly increased."

Budd reasoned that since automobile engineers had derailed the passenger train, they could be the ones to put it back on the tracks. As author David Morgan noted in *Diesels West!*, such thinking implied the use of a kind of automotive internal combustion power instead of the steam locomotive, upon which the industry had relied for a century.

One of Budd's first moves was to visit the Edward G Budd (no relation) Manufacturing Co in Philadelphia. The Budd Co had been the first to produce the all-steel automobile wheel, as well as the all-steel automobile body. Now it was ready to tackle passenger train construction.

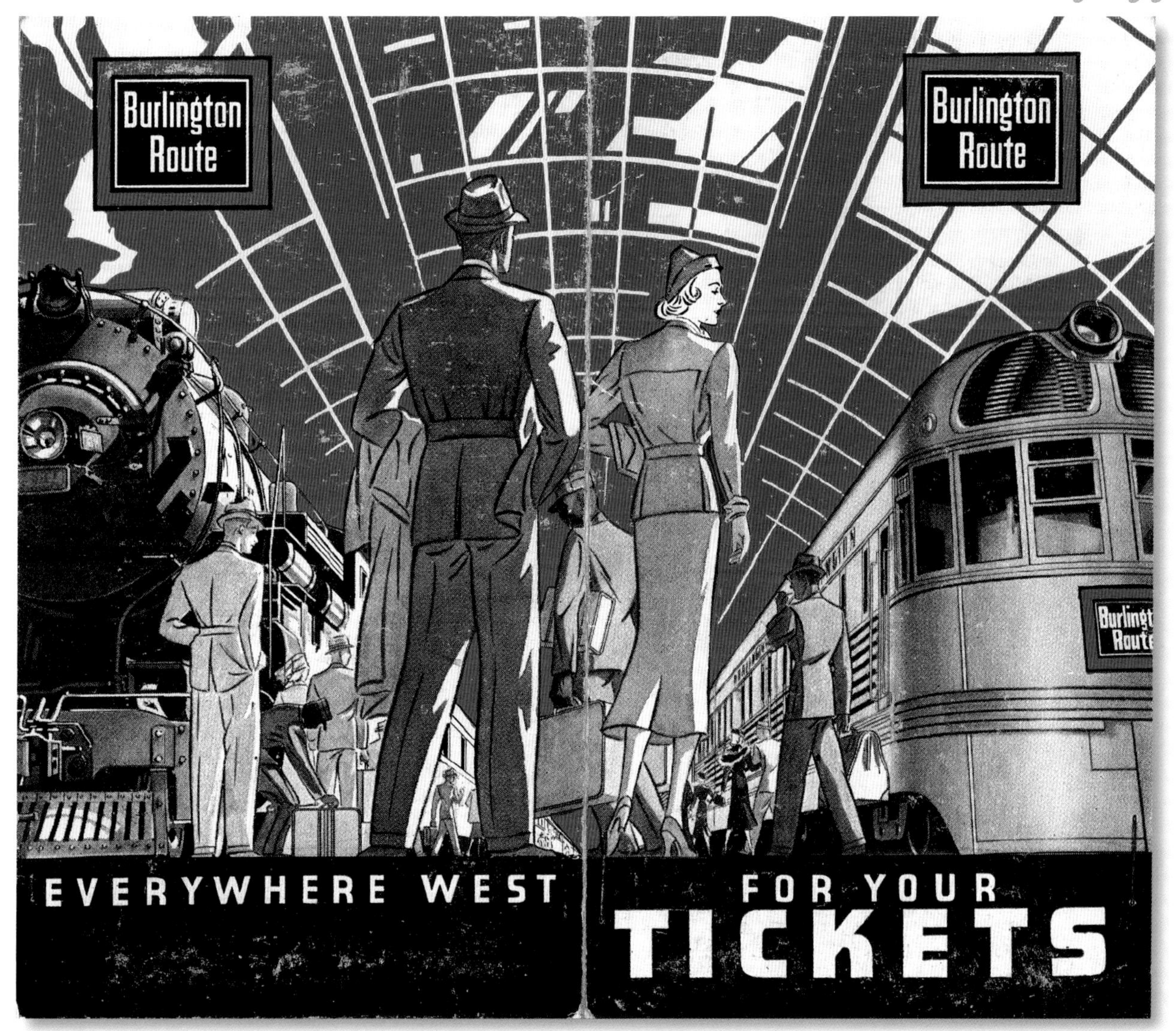

The front and rear of a contemporary ticket folder for the Pioneer Zephyr. Note the contrast between the streamlined sleekness of the Zephyr with the traditional steam locomotive on the left. (Owned by the author)

Spot-welded stainless steel

At the Budd Co plant, the Burlington president examined a test railcar body incorporating a number of radical innovations, including a petrol (gas) engine and rubber tyres. But Budd immediately dismissed these alterations as impractical. What caught his attention was the car's stainless steel construction. The benefits of the stainless steel railcar were obvious: the material was lightweight yet strong, and it lasted almost forever. The problem was nobody had been able to work out a way to build stainless-steel railcars in a practical operation.

Fortunately, the Burlington president's timing was just right. The Budd Co had just patented its newly devised 'Shotweld' process (electric spot-welding). The system allowed stainless steel to be joined without damaging its corrosion-resistant qualities. At the same time, it provided a joint stronger than the steel it held together. Ralph Budd knew a good product when he saw one. He decided this would be the material for the new train he set out to build to recapture the rail passenger market. On June 17, 1933, barely a year after he first stepped into office, Ralph Budd signed a contract with the Budd Co to construct a train out of stainless steel. The firm was given virtually free rein in the design of the train.

Now Budd needed a power unit for his stainless-steel train, to replace the outdated steam engine. The expertise of no less than three firms was brought to bear on the problem of building a two-cycle diesel engine to meet the demands for train engines with increased horsepower. The work was led by the General Motors Co, which was greatly aided by the company's 1930 acquisition of the Winton Engine Co and railcar builder Electro-Motive Co.

But it was Charles F Kettering, General Motors' vice president of research, and the engineering staff of Winton Engine Co, that led the project to a successful conclusion. In the late 1920s, Kettering began work on the two-cycle diesel engine, which eventually would become an integral part of the Pioneer Zephyr. His aim was to design a lightweight power unit with improved response and lower cost per horsepower than the available engines.

A major turning point occurred in 1933, when the Kettering team's eight-cylinder, 450kW (600hp), 8-201 engine, with a weight-to-power ratio of only 12kg per kilowatt (20 pounds per horsepower), was chosen to supply power to the Chevrolet exhibit at Chicago's Century of Progress Exposition. While visiting the fair, Ralph Budd came upon the display, and immediately decided that the light-weight diesel engine would provide the power for his new train. As Budd saw it, the diesel railroad was the railroad of the future—and if any company could put the diesel engine in a train, it was General Motors.

Record-breaking 1000 mile run

Only about a year later, on May 26, 1934, the Pioneer Zephyr No 9900 made its grand debut with a record-setting 1600km (1000mi) dawn-to-dusk run from Denver to Chicago in 13 hours. Appropriately enough, the Zephyr was named after the Greek god of the west winds.

The train was powered by a Winton 8-201A 450kW (600hp), two-cycle diesel engine (a revision of the model Ralph Budd saw at the fair), designed to travel at speeds of approximately 180km/h (110mph). Not only was the Pioneer Zephyr faster and lighter than its predecessors, it also reduced the Burlington's cost of passenger train operation.

The high operating standards of the world's first high speed diesel-propelled, stainless steel three-car train were matched by the Zephyr's painstaking interior furnishings. The first car held the diesel engine, engineer's cab, a 9.1m (30ft) railway post office, and space for baggage. The second car carried a larger baggage compartment, a buffet grill and, at the rear portion of the unit, a 5m (16ft) smoking section with seats for 20 passengers. Third and last was a 9.5m (31ft) compartment with seats for 40 persons, and a solarium-lounge with chairs for 12. Prior to the Pioneer Zephyr, the travelling public knew only ornate but gloomy railroad car interiors. All that changed with the Zephyr. Each individual compartment had a distinctive colour harmony coordinating wall colours, window drapes, upholstery, and floor covering.

Following the Zephyr's first historic run, it made another cross-country tour covering 222 cities, where it was received by some 2 million spectators. The train also was a hit attraction at Chicago's 1934 Century of Progress.

On Armistice Day 1934, the Zephyr went into regular passenger service, serving the West and Midwest until

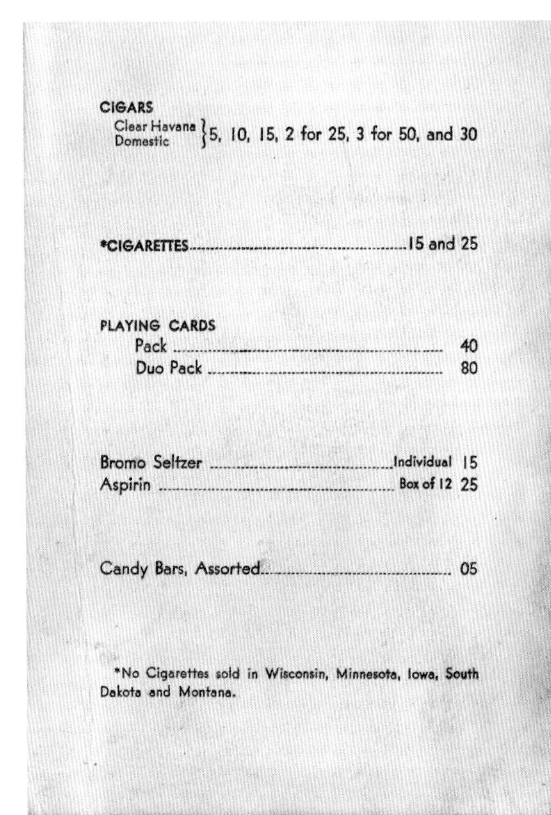

CIGARS
Clear Havana / Domestic } 5, 10, 15, 2 for 25, 3 for 50, and 30

*CIGARETTES .... 15 and 25

PLAYING CARDS
Pack 40
Duo Pack 80

Bromo Seltzer .... Individual 15
Aspirin .... Box of 12 25

Candy Bars, Assorted 05

*No Cigarettes sold in Wisconsin, Minnesota, Iowa, South Dakota and Montana.

The inside of a menu card for one of the Burlington trains. Twins of the Zephyr spread far and wide as the company capitalised on the success of the original. (Owned by the author)

A commemorative plate showing the Burlington Zephyr. A wide range of souvenirs featuring the train were available. (Owned by the author)

its retirement. But that was only the beginning – for the diesel and for the Pioneer Zephyr. The first million miles of the Pioneer Zephyr were celebrated on December 29, 1939. On April 10, 1944, the Zephyr was the guest of honour at its 10th birthday party at Lincoln, Nebraska. The granddaddy of diesel-powered streamlined trains celebrated its 20th anniversary of regular service at Quincy, Illinois. Finally, on May 26, 1960, the Pioneer Zephyr pulled up to its final destination just outside the East Pavilion of Chicago's Museum of Science and Industry, blowing the whistle on an unparalleled career that had spanned 26 years and some 5.1 million kilometres (3.2 million miles).

FASTEST IN STEAM

It's one of the oldest records in speed – the 125mph (200km/h) achieved by the steam locomotive Mallard in 1938

The best description of the record-breaking run of Mallard, still the world's fastest steam locomotive, was published in 1950 in a book called *Speed – the Book of Racing and Records*, published by Temple Press. Here it is, unabridged (and left in contemporaneous Imperial units). It makes me come out in goose-bumps to read it.

"There was no particular excitement at Doncaster locomotive shed on the morning of Sunday, July 3, 1938, as the beautiful blue stream-lined "Pacific" No. 4468 Mallard was being prepared for the road.

True, everyone knew she would reach 100mph that day; but with the day-by-day 90, 95 and 100mph attained by the Kings Cross-Newcastle-Edinburgh streamliners, the Silver Jubilee and the Coronation, the three-figure mark had lost much of its magic, even though it meant covering 146 feet every second with a heavy train.

Braking trials were the idea of the runs for that Sunday, as part of studies being made to give the celebrated streamline trains a shorter stopping distance from their "ceiling" speeds, for they needed 5,000 feet or more to come to rest from 95 or 100mph, and even then the brake blocks were known to have become red-hot.

The train was specially selected for the day, and consisted of the London and North Eastern Railway's dynamometer – or measuring – car and three twin-coaches from the spare train held for the London-Edinburgh "Coronation" service. Several of the railway's technical assistants and Westinghouse Brake Company's officials were the only passengers in the train, and Locomotive Inspector Jenkins was on the footplate along with the driver and fireman.

Approximately 100mph was needed for the brake tests; that is, a full 13mph below the L.N.E.R. top speed put up nearly three years before by Mallard's sister engine Silver Fox, so no records were likely. But a day or two before, when discussing the coming brake tests, Sir Nigel Gresley, the Chief Mechanical Engineer, had been heard to "wonder whether the 113mph could not be topped."

That was enough, in Gresley's day, for everyone concerned to have a go at it.

So Driver Duddington of Doncaster shed was inspanned; and Mallard, an exceptionally fast-running engine because of its double exhaust, was chosen from the list of 35 streamlined Pacifics. Duddington was known as a "speed merchant", ever ready to make up lost time by hard running over long distances, and well backed-up by Fireman Bray.

All L.N.E.R. top speeds in those golden years, when 90s were as common as blackberries, were attained on the gently-downhill stretch for 24 miles southwards to Peterborough. This was the best-known galloping ground on the whole of the East Coast route, and had been since the Flying Scotsman first started running in June 1862. Coming down from the north, there is a five-mile rise from Grantham to Stoke tunnel; then it is downhill through Essendine for 20 miles, and level for the last seven into Peterborough.

Speed and brake observations were made all the way south from Doncaster to Grantham; but any hidden ideas Duddington and his chiefs on the train might have had about records round Essendine way must have had a

MALLARD
No. 4468
CLASS
A4

Still the fastest steam locomotive in the world – Mallard. However, the run was not without problems, the locomotive suffering bearing damage as she was exerted well beyond her normal top speed. Photographed at the National Railway Museum in York, UK.

set-back here, for there was some of the usual Sunday track relaying in progress, and at half-past four in the afternoon speed through Grantham station could be no higher than 24mph

All drivers out for lost-time recovery or high speed down the bank to Peterborough like to pass Grantham as fast as they can, unless, indeed, their train is booked to stop there, so that they can rush the five-mile bank and pass over the top at something like the mile-a-minute rate, and have a chance of getting well up into the "eighties" before they slack off a bit for the Werrington water-troughs and the 25mph speed limit through Peterborough station.

But this hindrance was just the thing to put Duddington on his mettle. Now 1 in 200 may not seem much of a hill – you'd scarcely notice it on the road; but it's a different matter when there is 240 tons behind the tender to be hauled to the top as quickly as possible.

So Duddington put his regulator full open, set the valve gear at 40 per cent. cut-off, and Mallard fairly barked her way up through Great Ponton, and five miles on dived into Stoke tunnel at 69mph; Bray, meanwhile, having been piling seven or eight big shovelfuls of coal a minute on to the roaring fire.

One mile further on, just past the end of the tunnel and exactly 100 miles from London, is the little Stoke signal box, right at the top of the bank, and here the train was doing 75mph Still Duddington kept his regulator wide open and his cut-off at 40 per cent.; and the acceleration, which uphill from Grantham had been enough to raise the eyebrows, now became startling as the train got on to the 1 in 178 and 1 in 200 down grades. At the 99th post from London Mallard was up to 88mph, at the 98th post 97mph, and the train swept through Corby station, just by the 97th post, at 104mph

This was enough for brake trials. But Duddington had heard of Sir Nigel's "wonder," Bray was at the top of his firing form, and Jenkins was not the man to call "halt"; so on they went, the streamliner now riding with that easy roll and gentle airy feeling of a slight lift which all these wonderful A4-class "Pacific" engines take on at 100 to 105mph

By the time the 96th milepost was passed the train was up to 108mph, at the 95th post it was 111mph, and at the 94th had exceeded the British speed record of 114mph held since June 29, 1937, by the London Midland & Scottish Railway "Pacific" locomotive Coronation.

Here Duddington was experimenting with an even higher cut-off in Mallard's cylinders ; but within a mile – that is, within 32 seconds – he was back again to 40 per cent. because the extra steam couldn't get out of the cylinders quickly enough after it had done its work, and was scarcely increasing the speed.

Back to 40 per cent., up went the speed again. At the 93½ post it topped the two miles a minute mark; at the 92nd mile-post it was 122mph, and if Duddington didn't know the steam locomotive record speed of 124mph put up by a huge German streamlined 4-6-4 locomotive between Hamburg and Berlin, the men in the train did, and in the dynamometer car the excitement at having well and truly broken the British railway speed record was now being capped by "Will he break the world record?"

He did. Just before the 91st milepost speed reached 125mph; indeed, the rolling paper chart in the dynamometer car showed a tiny peak of 126mph, but not of long enough duration to be counted in a railway record, for if you are only one-fifth of a second out of dead accuracy over a distance of quarter of a mile at 120mph your speed reading is 3mph wrong.

Over the next mile, speed ranged at 124-125mph, the 6 feet 8 inch driving wheels making 81 revolutions every second, and each of the three pistons travelling 38 feet and stopping and starting 171 times each second.

That was the end, for steam was shut off and the brakes applied because of the approach to Essendine station with its slight curve and several sets of points and crossings, where something a bit under two miles a minute was needed for safety. So Essendine was passed at a mere 110, and then one of the brake tests was made, and in a mile the speed of Mallard and the seven coaches – weighing 400 tons in all – was brought down from 110 to 81mph; and at round about the 70 mark Mallard continued all the way until braking for the Peterborough stop.

Today her world record for a steam locomotive of 125mph, and of more than 120mph for three miles on end, still stands, and a long time is likely to pass before it is broken."

THE X-15

Fastest in the world

Capable of a speed more than twice as fast as the official current world record holder, the rocket-powered X-15 research aircraft achieved almost science fiction feats ... way back in the 1960s.

An unofficial motto of flight research of the 1940s and 1950s was 'higher and faster.' By the late 1950s, the last frontier of that goal was hypersonic flight (an incredible Mach 5+) to the edge of space. But to achieve this would require a huge leap in aeronautical technology, life support systems and flight planning.

Flies so high that pilots qualified as astronauts

The North American X-15 rocket plane was built to meet that challenge. It was designed to fly at speeds up to Mach 6, and altitudes up to 76km (250,000ft). The aircraft went on to reach a maximum speed of Mach 6.72 (7300km/h, or 4500mph) and a maximum altitude of 108km (354,200ft). Looking at it another way, Mach 6 is about one mile per second, and flight above only 80km (264,000ft) qualifies an Air Force pilot for astronaut wings ...

These winged aircraft speed and altitude records still stand – nearly 60 years after the aircraft first flew.

Three X-15s made 199 flights during a research program which lasted from 1960 through to 1968. The plane was air-launched by NASA's converted B-52 at 13,700m (45,000ft) and a speed of 800km/h (500mph). Generally, there were two types of flight profiles: high-speed, or high-altitude. High-speed flights were usually done below an altitude of 30,000m (100,000ft) with the X-15 flown as a conventional airplane using aerodynamic controls. High-altitude flights began with a steep, full-power climb to leave the atmosphere, followed by up to two minutes of 'coasting up' to the peak altitude, after which the engine was shut down. 'Weightless' flight would last for 2-5 minutes as the X-15 made a ballistic arc before re-entering the atmosphere.

Depending on the mission, the rocket engine provided thrust for the first 80 to 120 seconds of flight. The remainder of the normal 8-12 minute flight was without power, and ended in a 320km/h (200mph) glide landing. Because the nose landing wheel lacked steering and the main landing gear employed skids, the X-15 had to land on a dry lake-bed. The Rogers Dry Lake adjacent to Edwards and Dryden bases was the intended landing location for

An X-15 in the National Air and Space Museum, Washington DC, USA. Despite its age, it still looks extraordinary.

A rare photo of the X-15 at speed. Note the lower part of the tail-fin, which was jettisoned prior to landing. (Courtesy NASA)

all flights, numerous other lake-beds were selected in advance for emergency landings.

Although the number two aircraft was later modified, the basic X-15 was a single-seat, mid-wing monoplane designed to explore the areas of high aerodynamic heating rates, stability and control, pilot physiological phenomena, and other problems relating to hypersonic flight. Because the Reaction Motors Division of Thiokol Chemical Corporation did not have the throttleable XLR-99 engine ready for the early flights of the aircraft, the X-15 initially flew with two XLR-11 engines, producing a thrust of 7400kg (16,380lb). Once the XLR-99 was installed, the thrust became 26,000kg (57,000lb).

Different control systems for atmosphere and space

The X-15 used conventional aerodynamic controls for flight in the dense air of the usable atmosphere. The controls consisted of rudder surfaces on the vertical stabilisers to control yaw (movement of the nose left or right), and canted horizontal surfaces on the tail to control pitch (nose up and down) when moving in synchronisation, or roll when moved differentially. For flight in the thin air outside the Earth's atmosphere, the X-15 used a reaction control system.

A unique feature of the control system was the three control sticks in the cockpit. One was a centre stick, which controlled the airplane in pitch and roll as it would in a conventional aircraft; this was directly linked to another stick that was at the pilot's right side. The latter was operated by hand movement only, so the pilot's arm could remain fixed during high acceleration experienced during powered flight and re-entry. This was an essential feature, that enabled the pilot to maintain precise control for these conditions. The third control stick was located at the pilot's left, and was used to control the X-15 when it was above the atmosphere. This stick actuated the reaction jets, which utilised a form of superheated steam, generated from the decomposition of hydrogen peroxide (H_2O_2). The reaction thrust was produced by small rocket motors located in the nose, for pitch and yaw control, and within the wings for roll control.

Nickel-chrome alloy skin

The outer skin of the X-15 consisted of a nickel-chrome alloy called Inconel X, employed in a heat-sink structure to withstand the results of aerodynamic heating when the aircraft was flying within the atmosphere. Temperatures as high as 690°C (1280°F) were experienced on the skin of the aircraft – enough to make steel glow red-hot. Considerable weight was saved through the use of titanium in parts of the internal structure not subject to high temperatures. The cabin was made of aluminium, and was isolated from the outer structure to keep it cool.

The X-15s were roughly 15.2m (50ft) long, with a 6.7m (22ft) wing span. The wedge-shaped vertical tail was 4m (13ft) high. Because the lower vertical tail extended below the landing skids when they were deployed, a part of the lower vertical tail was jettisoned just before landing, and recovered by a parachute.

Although by 1955 the United States had eight years' experience with aircraft rocket engines, one with the required 50,000-lb thrust was a big advance over any previously used for that purpose. In fact, missiles had

THE RECORD?

Despite routinely exceeding speeds of Mach 6, at the time of the X-15 flights, the official world speed record for winged aircraft was just Mach 3, held by the Soviet E166. So why wasn't the X-15's speed ratified as a world record? One reason is that the X-15 could not take off under its own power – a feat achievable by the pictured 30-year-old SR71 Blackbird, the current nominated record holder with a speed of (just!) 2200mph (3500km/h) – less than half the speed of the incredible X-15. This SR-71 was photographed at the Steven F Udvar-Hazy Museum, Virginia, USA.

The X-15 was air-launched by NASA's converted B-52 at 45,000 feet and a speed of 500mph. (Courtesy NASA)

provided the only experience with large rocket engines. But the X-15 couldn't become a one-shot operation – its engine would have to be an aircraft engine, capable of variable thrust over at least 50 per cent of the thrust range and having other normal cockpit control features, such as restarting.

Subsequently, Reaction Motors, Inc was selected to develop what became the XLR-99 rocket engine. The technical requirements contained a specification that "any single malfunction in either engine or propulsion system should not create a condition which would be hazardous to the pilot."

The engine burned a mixture of anhydrous ammonia (NH_3) and liquefied oxygen (LO_2). These propellants posed a few handling problems, because of the corrosive properties of ammonia and the low temperatures of liquid oxygen, which boils at -182°C (-297°F). Since the propellant tanks were an integral part of the airplane structure, temperature extremes between structure close to the liquefied oxygen tank and surrounding structure exerted a major influence on thermal stresses and design. The liquefied oxygen tank had a capacity of 3800 litres (1003 gallons); the ammonia tank, 5500 litres (1445 gallons). This gave a burning time of 85 seconds at full thrust. An important feature of the X-15's liquefied oxygen system was the need for replenishing it after take-off, because of the large amount lost through boil-off during the climb to launch altitude aboard the B-52. This topping-off took place continuously, under control of a B-52 crewman, from tanks within the B-52, which had a capacity 1.5 times that of the X-15.

Temperature extremes

Besides engine propellants, the X-15 also carried vast quantities of hydrogen peroxide (H_2O_2), liquefied nitrogen (-190°C/-310°F), gaseous nitrogen, and gaseous helium (-150°C/-240°F), used to operate various subsystems. With such large amounts of super-cold liquids flowing within the airplane, its internal components needed protection from freezing, not high temperature!

A major subsystem was the air conditioning unit, which protected the pilot and instrumentation from the effects of heating and also cooled the auxiliary-power system. It operated from liquid nitrogen, and in addition to cooling, pressurised the cockpit, instrument compartment, pilot's suit, hydraulic reservoir, and canopy seal.

The X-15 had its share of emergency landings and accidents, but only two produced serious injuries or death. On November 9, 1962, Jack McKay experienced an engine failure and landed at Mud Lake, Nevada. The landing gear collapsed, flipping him and the aircraft on its back. Although he sufficiently recovered from his injuries to fly again, he eventually had to retire because of them. On November 15, 1967, on Michael Adams' seventh flight, he entered a spin from which he was able to recover but could not bring it out of an inverted dive because of a technical problem with the adaptive flight control system. Sadly, he died in the resultant crash of the X-15 number three.

In terms of its advancement in aircraft and space technology, the X-15 made an enormous impact. And, although rarely mentioned in the record books, the X-15 remains the fastest, highest-flying plane ever to take to the air ...

The mothership overflies an X-15. Note the rear skids on which the aircraft landed – it could land only on dry lake beds. (Courtesy NASA)

BEATING THE SOUND BARRIER

Perhaps the most famous of the 'X' series of experimental aircraft was the first – the Bell X-1. It was in this aircraft that Chuck Yeager officially exceeded the speed of sound, travelling at Mach 1.06 (700mph/1100km/h). The aircraft, named Glamorous Glennis after his wife, is surprisingly petite and beautiful in the metal. It was photographed at the National Air and Space Museum, Washington DC, USA.

The Golden Arrow beat the Land Speed Record in 1929. It took a radically different approach to the 1000hp Sunbeam, shown earlier in this chapter. Rather than use a full width body, the streamlining aimed to minimise frontal area, an approach that this photo shows well. It used a single 24-litre 'broad arrow' 12-cylinder Napier Lion series VIIa liquid-cooled engine. Driven by H O D Segrave, at Daytona Beach, it achieved an average speed over the flying mile of 231.446mph (372.459km/h). The car is pictured at the National Motor Museum at Beaulieu, UK.

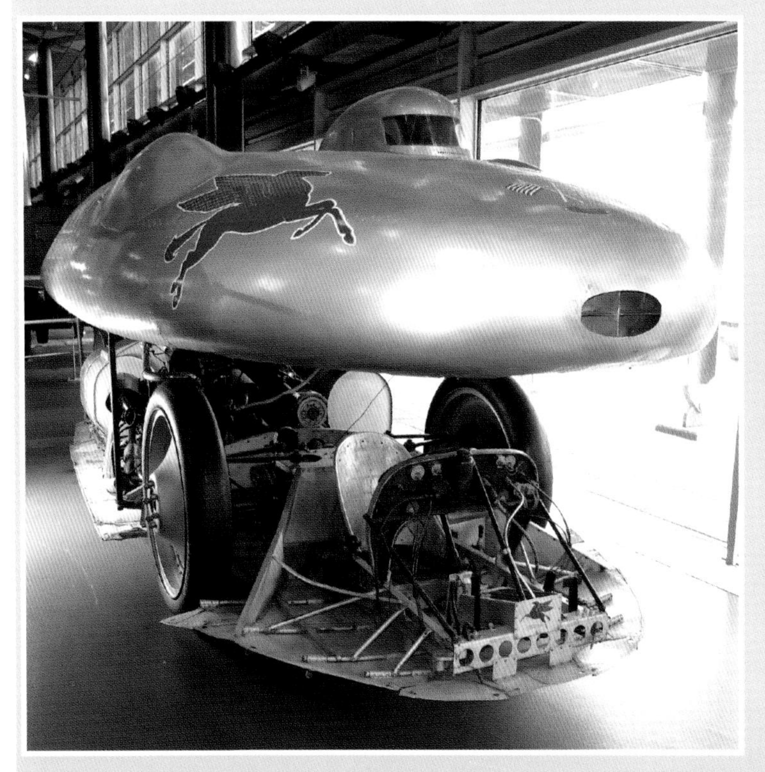

One of the most extraordinary Land Speed Record breakers ever was the Railton Special – later renamed the Mobil Railton Special after one of its sponsors. The Railton, driven by John Cobb, took the record in 1938 (350.2mph/ 563.566km/h), 1939 (369.74mph/595.05km/h) and then again after World War II in 1947 (394.196mph/634.397km/h). The car used an S-shaped chassis spine alongside which were mounted two 23.9 litre Napier Lion engines, one driving the front wheels and the other, the rear wheels. Streamlining reached a new level of proficiency for Land Speed Record cars. Here it is pictured with its lightweight body lifted above the chassis – an approach taken whenever work needed to be done on the car. Photographed at Thinktank, Birmingham Science Museum, UK.

The Mercedes Type 80 was prepared in the late 1930s for an attempt on the world Land Speed Record. However, World War II intervened, and the car never ran. Josef Mikcl did the aero design that was superbly streamlined, with a claimed drag coefficient of just 0.18. In addition to the all-enveloping low-drag body, thin horizontal wing extensions each side of the body were used to reduce lift. The vehicle was to use a cutting-edge aircraft engine – the Daimler Benz DB600 series V12. Power was estimated at being between 2100-2600kW (2830-3500hp). A realistic potential speed for the T80 was 650km/h (400mph). Photographed at the Mercedes-Benz Museum, Stuttgart, Germany.

Craig Breedlove's Spirit of America showed a new approach to Land Speed Record cars. It had only three wheels, and those wheels were not driven but instead free-wheeled as the car was propelled solely by jet thrust. In fact, initially its 1963 speed of 407.447mph (655.722km/h) was disallowed as the official record, but common sense soon reigned, and the car was then heralded as the fastest in the world. Considering the low resources with which the machine was developed and built, the car looks stunningly professional in the metal. It was photographed at the Museum of Science and Industry, Chicago, USA.

Donald Campbell's Bluebird is one of the most fascinating Land Speed Record Cars every built. Produced with the weight of British industry behind it, the very advanced jet-powered (but wheel-driven) car set the record at Lake Eyre in South Australia. Its record run was 403.10mph (648.73km/h) – faster than the previous wheel-driven record set by John Cobb in the Railton Mobil Special but slower than the thrust-propelled Spirit of America. It was also quite a lot slower than the speed the car appeared to be capable of – poor salt conditions being responsible in part for that. The car is pictured at the National Motor Museum at Beaulieu, UK.

Goldenrod was developed by brothers Bob and Bill Summers. It held the wheel-driven Land Speed Record from 1965-1991. The car was powered by no less than four fuel injected Chrysler Hemi engines, mounted inline and creating a total output of 2400hp (1800kW). The car had a very low coefficient of drag, and this combined with a tiny frontal area, allowed the car to set the record at 409.277mph (658.64km/h). To run faster than Bluebird on a relatively tiny budget was an enormous credit to the two brothers. Photographed at the Henry Ford museum, Michigan, USA.

Thrust2 held the Land Speed Record from 1983 to 1997. It broke the record at 633.468mph (1,019.468 km/h). The machine was powered by a single Rolls-Royce Avon jet engine and had a configuration similar to the 1960s jet-powered land speed record cars of the American, Art Arfons. Thrust2 was driven by Richard Noble. The car is brutish and ugly; it was photographed in the Coventry Transport Museum, Coventry, UK.

ThrustSSC is the current Land Speed Record holder – the fastest car in the world. On October 15, 1997, it set a record speed of 1,228km/h (763mph), becoming the first car to officially break the sound barrier. Driven by Andy Green, it was powered by two afterburning Rolls-Royce Spey turbofan engines. The record-breaking runs were made at Black Rock Desert, Nevada, USA. ThrustSSC is in the Coventry Transport Museum, Coventry, UK. In the metal it is one of the most impressive of all the Land Speed Record cars – as befits its superlative performance.

Acknowledgement: Some of the text of this chapter is based on material drawn from the American Society of Mechanical Engineers (Pioneer Zephyr) and NASA (X-15) and is used with permission. The Pioneer Zephyr is a Historic Mechanical Engineering Landmark.

Locations of photographs

Attraction	Location	Comments, duration of visit, and author's rating
Aeromuseum	Gothenburg, Sweden	Fascinating underground bunker containing an aviation museum with lots of hands-on activities. (2.5 hours, 9/10)
Boneyard at the Davis-Monthan Air Force	Tucson, Arizona, USA	Very detailed guided tour; you are not allowed to get off the bus. (1 hour, 8/10)
Bradbury Science Museum	Los Alamos, New Mexico, USA	Disappointing – too much information and too few artefacts. (30 minutes, 4/10)
British Motor Museum	Gaydon, UK	Ex-British Leyland collection. Some extraordinary prototypes and rarities. (Half a day, 9/10)
Brooklyn Bridge	New York City, New York, USA	Wonderful bridge to walk across. (2 hours, 8/10)
Coventry Transport Museum	Coventry, UK	Absolutely superb museum that includes an excellent line-up of cars and bicycles. (Half a day, 9.5/10)
Deutsches Museum	Munich, Germany	Unbelievably good – probably the best in the world. (2 days, 11/10 – yes, 11/10!)
Deutsche Technikmuseum	Berlin, Germany	Outstanding museum in all things technical. (Spend a full day, 9/10)
Discovery Museum	Newcastle upon Tyne, UK	Contains Turbinia – the first steam turbine boat. (1 hour, Turbinia: 8.5/10; rest of museum: 6/10)
Erasmus bridge	Rotterdam, the Netherlands	Lovely bridge, good dusk walk. (1 hour, 8/10)
Falkirk Wheel	Falkirk, Scotland, UK	Marvellously artistic engineering structure. (2 hours, 8.5/10)
Firth of Forth railway bridge	Firth of Forth, near Edinburgh, Scotland, UK	My favourite bridge in the whole world! Truly magnificent structure. No access to bridge. (2 hours, 10/10)
Fleet Air Arm Museum	Somerset, UK	High quality aviation museum. Unexpectedly good aircraft carrier experience. (2-3 hours, 8/10)
Future of Flight Aviation Center	Seattle, Washington, USA.	Relatively small display, but worth looking at if present for the Boeing factory tour. (1 hour, museum: 7/10; factory tour: 9/10)
Golden Gate Bridge	San Francisco, California, USA	Pedestrian access is good, spectacular bridge and spectacular view. (2 hours, 8/10)
Haynes International Motor Museum	Somerset, UK	Really outstanding collection of a huge number of cars. Much under-rated. (2-3 hours – perhaps more if you want to take lots of photographs, 9/10)
Helicopter Museum	Weston-super-Mare, UK	Excellent exhibits but more space is needed. (1-2 hours, 8/10)
Henry Ford museum	Michigan, USA	Excellent technology museum, especially in mass production and resulting societal changes. Smart choice of cars on display. (Spend a full day, 9/10)
Hoover dam and power station	Nevada / Arizona, USA	Extraordinary engineering feat with unexpected style. Full tour is expensive but is the one to go on. (4 hours, 9.5/10)
Hovercraft Museum	Lee-on-the-Solent, UK.	Fantastic collection of supremely wonderful craft and associated paraphernalia. (Half a day, 10/10)
Iron Bridge and Museum of Iron	Iron Bridge and surrounding areas, UK	Fascinating insight into first uses of iron in industrial, civil engineering and domestic use. (3 hours, 9/10)
Louwman Museum	The Hague, Netherlands	Huge range of cars but could be better curated. (3 hours. 8/10)

Attraction	Location	Comments, duration of visit, and author's rating
Mercedes-Benz Museum	Stuttgart, Germany	Absolutely fantastic car museum. (Spend a full day, 10/10)
Mittelbau-Dora Concentration Camp Memorial	Near Nordhausen, Germany	Extraordinary place – visit the museum, the site of the camp and the underground tunnels in which the V2 were built. (Half a day, 9/10)
Museum of Science and Industry	Chicago, USA	Dumbed-down museum but with three major attractions – Zephyr train, Spirit of America LSR car, U505 WWII German submarine; expensive entry charges and parking. (2 hours. Overall: 3/10; nominated attractions: 9/10)
National Air and Space Museum	Washington DC, USA	Extraordinary museum, brilliantly curated with a collection second to none. Free entry. (Spend a full day, or two, if really interested in space and aviation, 10/10)
National Atomic Testing Museum	Las Vegas, Nevada, USA	Excellent collection that is well organised. First half of museum is outstanding. (3 hours, 8.5/10)
National Motor Museum	Beaulieu, UK	Outstanding collection of cars, especially Land Speed Record and motorsport machines. Expensive admission. (4 hours. 9/10)
National Museum of Flight	East Fortune, Scotland, UK	Significant aircraft and located on a historic airport. (5 hours, 9/10)
National Museum of Nuclear Science & History	Albuquerque, New Mexico, USA	Not quite as well organised as the National Atomic Testing Museum but has a better collection of artefacts. (3 hours. 8.5/10)
National Postal Museum	Washington DC, USA	Well thought out, small museum that can change your mind about stamps and postage! Free entry. (2 hours, 8/10)
National Railway Museum	York, UK	One of the best rail museums in the world. Incredible rolling stock and artefacts. (Full day, 10/10)
New Mexico Museum of Space History	Alamogordo, New Mexico, USA	Excellent space museum with mostly good signage and artefacts. (3 hours, 8/10)
Otowi Suspension Bridge	Santa Fe County, New Mexico, USA	Worth stopping to look at if driving to Los Alamos – classic engineering structure. (15 minutes, 7.5/10)
Peenemünde Historical Technical Museum	Peenemünde, Germany	Superbly curated museum at the birthplace of modern rocketry. (Spend half a day – or a full day if you have a car and want to explore more of the site, 9/10)
Petersen Automotive Museum	Los Angeles, USA.	High quality and diverse range of cars. 'Vault' tour in basement prohibits photography. (4 hours (inc Vault), 8.5/10)
Pierce-Arrow Museum	Buffalo, New York, USA	Small but excellent, with some glorious cars and a good variety of automobilia. (2 hours. 8/10)
Pima Air & Space Museum	Tucson, Arizona, USA	Outstanding collection of (mostly) military aircraft. (Full day (if including Boneyard tour), 9/10)
Science Museum	London, UK	Fantastic technology museum – we went back three times. (Full day. 10/10)
Steven F. Udvar-Hazy Museum	Virginia, USA	Extraordinary museum, brilliantly curated with a collection of full-size aircraft second to none, free entry. (Spend a full day, or 2 days if really interested in space and aviation, 10/10)
Technik Museum Sinsheim	Sinsheim, Germany	One of the world's best technical museums, with an emphasis on cars. (Spend a full day, 10/10)
Technik Museum Speyer	Speyer, Germany	Eclectic and wonderfully stimulating technology museum. (Full day, 10/10)
Thinktank, Birmingham Science Museum	Birmingham, UK	Fantastic Mobil Railton Special LSR car, other good exhibits. (2 hours, 8.5/10)
Titan Missile Museum	Arizona, USA	One of the most extraordinary technical places to visit in the world. (3 hours, 10/10)
Tower Bridge	London, UK	Unusual and interesting bridge, with tour available of old steam operating machinery. (2 hours, 8/10)
York Cold War Nuclear Bunker	York, UK	Fascinating Cold War nuclear shelter in well-preserved condition. Excellent tour. (2 hours. 10/10)
USS Bowfin Submarine Museum	Honolulu, Hawaii, USA	Excellent submarine access and interesting museum. (2 hours, 7.5/10)
White Sands Missile Range	New Mexico, USA	Excellent outdoor and indoor displays. (3 hours, 9/10)
Wright Brothers National Memorial	Kill Devil Hills, North Carolina, USA	Extraordinarily moving place to visit, good smaller museum. (3 hours, 9/10)
Zeppelin Museum	Friedrichshafen, Germany	Fascinating for anyone interested in airships. (3 hours, 8.5/10)

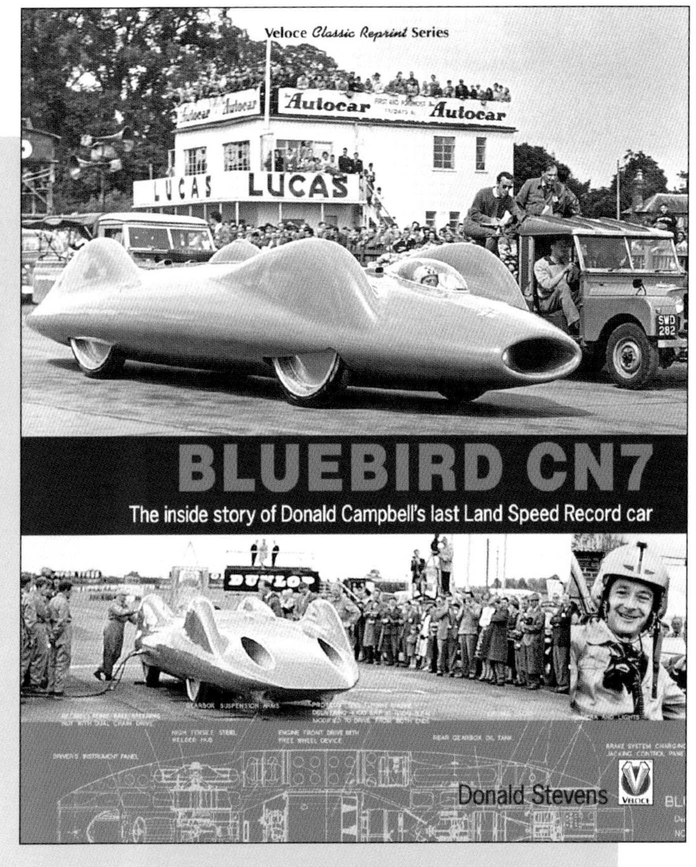

Veloce Classic Reprint Series
Autocar
LUCAS
BLUEBIRD CN7
The inside story of Donald Campbell's last Land Speed Record car
DUNLOP
Donald Stevens
VELOCE

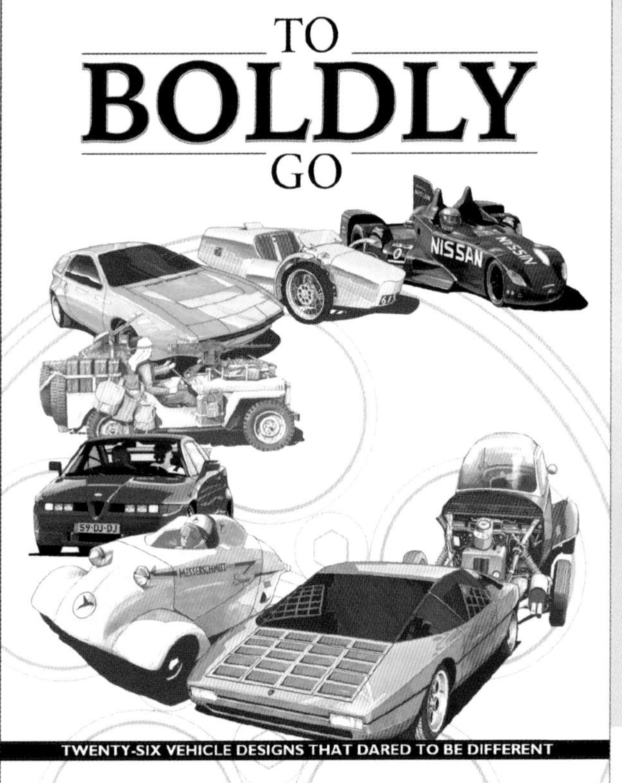

TO
BOLDLY
GO
NISSAN
TWENTY-SIX VEHICLE DESIGNS THAT DARED TO BE DIFFERENT
GRAHAM HULL

More books of interest from Veloce Publishing ...

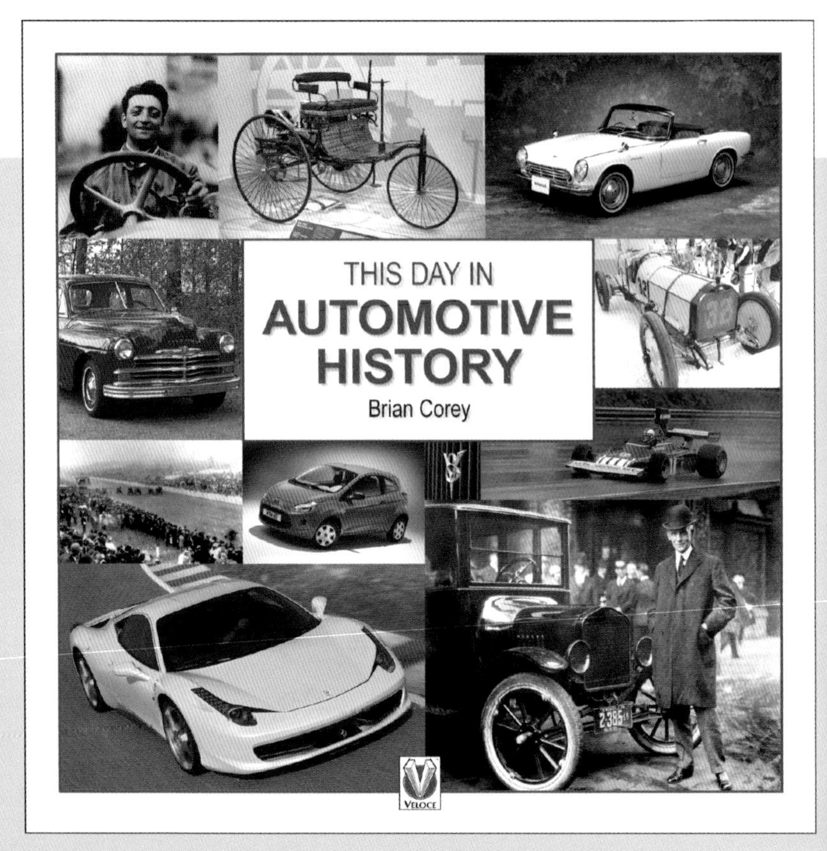

With a true tale for every day of the year, this book tells of the first self-propelled vehicles of the 1700s, the origins of Ford, FIAT and Jaguar, about the tragedies that have altered the daily commute, and about some of history's most important cars, such as the Porsche 356, Chevrolet Corvette, Volkswagen Beetle, Ford Model T, and much more.

ISBN: 978-1-787110-68-7
Paperback • 25x25cm
• 196 pages • 730 pictures

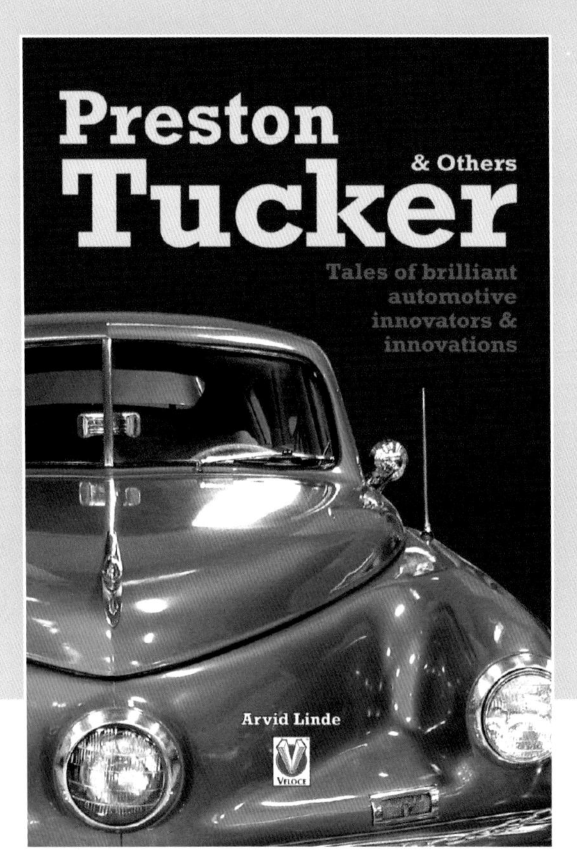

An automotive book that is guaranteed to surprise, intrigue, entertain and inform, whether the reader is an automotive expert or a complete novice; altogether a memorable, eye-opening journey through our automotive heritage. This book celebrates those pioneering individuals with a radical, non-conformist approach to car design, from Preston Tucker and his Torpedo to Guy Negre and his zero-emission dream.
Lavishly illustrated and intensively researched, this is also the story of the milestone inventions that have shaped today's cars and automotive landscape.

ISBN: 978-1-845840-17-4
Paperback • 22.5x15.2cm • 160 pages • 165 colour and b&w pictures

For more information and price details visit www.veloce.co.uk • email: info@veloce.co.uk • Tel: +44(0)1305 260068

Also by Julian Edgar:

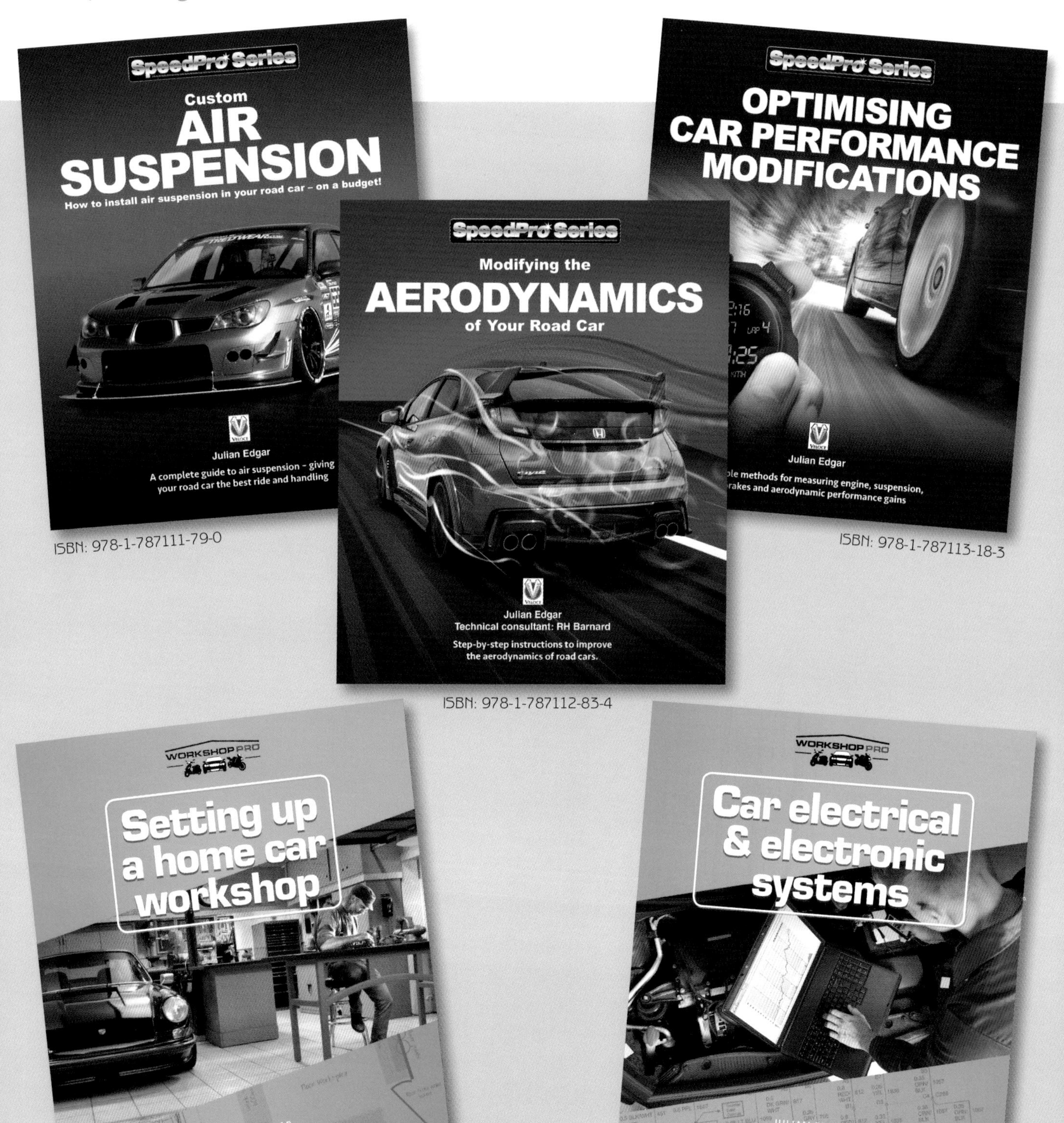

ISBN: 978-1-787111-79-0

ISBN: 978-1-787113-18-3

ISBN: 978-1-787112-83-4

ISBN: 978-1-787112-08-7

ISBN: 978-1-787112-81-0

Index